LIDDELL HART

By the same author

VICTORIAN MILITARY CAMPAIGNS (ed.) (Hutchinson)

THE VICTORIAN ARMY AND THE STAFF COLLEGE, 1854–1914 (Eyre Methuen)

CHIEF OF STAFF: THE DIARIES OF LIEUTENANT-GENERAL SIR HENRY POWNALL, 1933–44 (ed.) (2 vols, Leo Cooper)

FRANCE AND BELGIUM 1939–40 (Davis-Poynter)

LIDDELL HART
A Study of his Military Thought

Brian Bond

RUTGERS UNIVERSITY PRESS
NEW BRUNSWICK, NEW JERSEY

First published in the United States of America by
Rutgers University Press, 1977

A condensed version of Chapter 9 first appeared in *R.U.S.I. Journal* (June 1976) and of Chapter 8 in *Military Affairs* (February 1977)

Library of Congress Catalog Card Number 77–79718
ISBN 0–8135–0846–0

First published in Great Britain by
Cassell & Company Limited, 1977

Manufactured in Great Britain

Contents

Author's Note

My thanks are first and foremost due to Lady Liddell Hart for permission to undertake this study with unrestricted access to Sir Basil Liddell Hart's Papers. She has also most generously allowed me to quote extensively from Sir Basil's books in which she holds the copyright. Her generosity in other respects is well known to the many visitors to the archives at States House, Medmenham over the years, and I have benefited more than most. More specifically, as a close neighbour, I have enjoyed the inestimable advantages of being allowed to work in the archives at virtually any time that suited me, and to borrow any books that I needed from the splendid library. I should like to take this opportunity to express once again my gratitude for these privileges, which I have enjoyed ever since 1959.

I am also grateful to the Centre for Military Archives, King's College, London, and particularly to the archivist Miss Julia Sheppard, for allowing me to quote extensively from the Liddell Hart Papers which belong to the Centre, though still located at States House, Medmenham, Bucks at the time of writing.

I was fortunate in that the period of my most intensive research coincided with Stephen Brooks' final year as archivist at States House. Whereas I had known Sir Basil extremely well for over a decade, Stephen had never met him and consequently derived his impressions largely from the documents, which he catalogued with remarkable skill. This difference in backgrounds provided a focal point for our numerous discussions and served to clarify my views on various issues. Stephen also read my book in draft and made many valuable suggestions; we found ourselves in complete agreement on nearly every point.

I am also very grateful to Michael Howard for suggesting a way of approaching the subject; for his comments on the greater part of the book in draft; and not least for his general encouragement. Individual chapters were read by Dr R. J. O'Neill, Mr Richard Ogorkiewicz, Major Charles Messenger and Dr Martin van Crefeld. All my readers made helpful corrections and suggestions but I alone am responsible for any blemishes which remain.

My thanks are due to the Social Science Research Council for a travel grant which enabled me to visit Israel in September 1975. Mr Chaim Bar-On very kindly arranged my interviews and accommodation, and Mrs Ruth Connell Robertson and Dr Uri Bialer both helped greatly to make my short stay enjoyable.

While the book was in draft I read papers based on parts of it to graduate seminars at Reading and Sussex Universities, and at the Institute of Historical Research at the University of London. I am appreciative of the many constructive comments made at these meetings, and of the tolerance of my own graduate students during my long preoccupation with this subject.

My thanks are due to many others who helped in a great variety of ways, including (with apologies for any inadvertent omissions) Professor Norman Gibbs, Martin Gilbert, Klaus Gütig, Dr Paul Kennedy, General Haim Laskov, Adrian Liddell Hart, Ronald Lewin, Colin Lovelace, Lord Paget (formerly Sir Reginald Paget Q.C.), Dr R. A. C. Parker, Miss Nancy Scammell, Brigadier A. J. Trythall, Professor Yigael Yadin and Sir Edgar Williams.

One limitation I share with the subject of this book is that I cannot type. Consequently I am particularly grateful to Mrs Edna Robinson for retyping some of the chapters, and crucially indebted to my wife Madeleine for typing and retyping the bulk of the book from my untidy original drafts.

Brian Bond
Medmenham, Bucks
November 1976

Acknowledgements

My grateful thanks are due to the following for the use of unpublished material: Mrs Evelyn Arthur (General Sir John Burnett-Stuart's unpublished memoirs and letters to Liddell Hart); Brigid Brophy and Kate Levey (John Brophy's letters to Liddell Hart); David Higham Associates Ltd (Major-General J. F. C. Fuller's letters to Liddell Hart); Lady Pakenham-Walsh by the courtesy of Leo Cooper (diaries of Major-General R. P. Pakenham-Walsh); and Miss Nancy Scammell (Colonel J. M. Scammell's letters to Liddell Hart and Spenser Wilkinson).

For published material I am much indebted to Lady Liddell Hart for the following works by Sir Basil Liddell Hart: *Memoirs* (Cassell: Putnam), *Europe in Arms* (Faber: Random House), *The Defence of Britain* (Faber: Random House), *The Revolution in Warfare* (Faber: Yale U.P.), *Defence of the West* (Cassell: Morrow), *Deterrent or Defence* (Stevens: Praeger), *Strategy: the Indirect Approach* (Faber: Praeger), and other books by the same author which I have quoted less extensively; to Willoughby Pownall-Gray for again allowing me to quote from my edition of the Pownall Diaries *Chief of Staff* vol. I (Leo Cooper: Shoe String Press); to Michael Howard for quotations from his Neale Lecture *The British Way in Warfare: a Reappraisal* (Cape); to the Trustees of T. E. Lawrence's letters for *T. E. Lawrence to His Biographers* (Doubleday: Cassell); and to A. M. Heath & Co. Ltd on behalf of Mrs Sonia Brownell Orwell for *Collected Essays, Journalism and Letters of George Orwell* vol. II (Secker & Warburg: Harcourt Brace).

I should further like to acknowledge the following publications, which have all been extremely useful, though I have not quoted from them extensively: Jay Luvaas *The Education of an Army* (University of Chicago: Cassell); R. Macleod and D. Kelly (eds) *The Ironside Diaries 1937–1940* (Constable: McKay); Adam Roberts (ed.) *The Strategy of Civilian Defence* (Faber: Stackpole); H. Guderian *Panzer Leader* (Joseph: Dutton); R. M. Ogorkiewicz *Armour* (Stevens: Praeger); E. Luttwak and D.

Horowitz *The Israeli Army* (Allen Lane: Harper); A. J. P. Taylor (ed.) *Lloyd George: Twelve Essays* (Hamish Hamilton: Atheneum); and R. T. Paget *Manstein* (Collins).

Introduction

Academic historians are taught to avoid using the first person singular, but since an important motivation for this study was a close association with its subject I should perhaps begin by saying something about my relationship with Captain Sir Basil Liddell Hart.

I had the good fortune to meet him soon after he came to live at States House in my home village of Medmenham, Bucks in December 1958. I was then in my final year at Oxford reading Modern History and had already chosen the military history option for my special subject. By a happy coincidence the last two books I had read before meeting 'The Captain'—as I came to call him—were his *Ghost of Napoleon* and *Strategy: the Indirect Approach*, and since I had found both very stimulating I naturally got off to a good start with him. This would not have been difficult anyway because he was eager to help any student who showed promise. He at once displayed great interest in my work, devoting more time and attention to reading and marking my recent undergraduate essays than they deserved, gave me signed copies of his books and allowed me to use his incomparable library—a privilege from which I have benefited ever since. As Finals approached, he thought I was overworking and under-eating and with characteristic generosity gave me £5 to spend on steaks—a sum which in 1959 would have paid for half-a-dozen meals at least.

Basil's most important 'gift' to me at this time, however, was that he encouraged me to believe that I was good enough to make a career as a military historian. This was particularly valuable to me since I was rather lacking in self-confidence academically and my ambition to go on to do postgraduate research had been discouraged by my College tutors. If anything, his confidence in my ability was even more important when I was a research student without a grant, under the usual emotional pressures of that age, and with very dim prospects of a job since then, as now, academic posts in military history were extremely rare. He went to a lot of trouble in persuading me to 'stick it out' and in helping me to take the all-important first step on the academic ladder.

As I suggested earlier, I was by no means the only young man whom Liddell Hart went out of his way to help: some of the others are represented in the 'Young Rogues' gallery of signed photographs at States House. But I was particularly fortunate in living in the same village in that I could see him nearly every day, and at any odd hour that suited him—sometimes from 9 p.m.–11 p.m. before he returned to another late-night spell of writing. In return I helped rearrange his library and (with Professor Jay Luvaas) began the long process of sorting out his Papers, which were then (1961) in unimaginable confusion. I also acted as his unofficial research assistant, looking up information, reading books which he was too busy to do more than glance at, and checking his later books at various stages of typescript and proof.

Thus I came to know and admire Liddell Hart as a generous, warm-hearted friend and mentor before giving much thought to the validity of his theories or the extent of his influence; indeed I tended to a large extent to accept his own self-evaluation uncritically, helped by the fact that for several years I specialized in the Victorian period in which he laid no claims to expertise. Consequently I was never puzzled, like some of his friends who knew him first through his publications, by the divergence between the *persona* of the professional pundit and the private individual. This point was well expressed by Ronald Lewin in a commemorative radio broadcast:

> From the books you certainly didn't get the impression of warmth: here, you felt, was a forthright writer, a man who didn't suffer fools gladly, a man who knew his own mind—not really you felt a lovable man ... But of course when you met him the whole of that went by the board. Immediately you realized that here was a warm, generous, humorous person, with an enormous range of interests.

Michael Howard similarly wrote that 'Scores if not hundreds of students and disciples were bound to this implacable and loving master'.[1] The vast number, wide time-span and thickness of his correspondence files bear witness to his genius for making and retaining friendships with a remarkable variety of people.

It must be emphasized that Liddell Hart relished intellectual discussion and controversy both in conversation and in print. He liked people to stand up to him in argument, and in my experience was remarkably tolerant towards divergent views irrespective of the age or status of the person holding them. Being unashamedly egocentric, he liked to win a point and make converts to his way of thinking, but he would accept 'failure' philosophically provided the usual courtesies were maintained. Again his correspondence files prove that he rarely broke off a relationship out of pique, impatience or intolerance, contrary to what one might imagine from the *enfant terrible* whose arrogant articles outraged some of the less progressive generals in the 1920s.[2]

Since Liddell Hart was a tireless advocate of independent judgement in the pursuit of truth, he was probably prepared to face the fact that even his keenest pupils ('loyal' was not one of his favoured terms of praise) would disagree with him on some important points. In the last years of his life I was already uncomfortably aware that I did not see eye to eye with him on certain issues on which his mind was irrevocably made up. For example, in editing the diaries of Lieutenant-General Sir Henry Pownall[3] (who served on the secretariat of the Committee of Imperial Defence from 1933 to 1936 and was Director of Military Operations and Intelligence at the War Office in 1938–39) I came to the conclusion that, though less advanced than Liddell Hart in their thinking on mechanization, Pownall and his colleagues on the General Staff were for the most part intelligent professional officers wrestling with intractable strategic and financial problems which had not been fully and fairly presented by the critical journalist. For Liddell Hart, on the other hand, the myopic, obstructive character of the General Staff in the 1930s had become an article of faith which it would have been unreasonable to expect him to discard late in life.

When Liddell Hart died suddenly on 29 January 1970 we had just begun to collaborate on a volume of quotations to illustrate the outstanding military thinkers from Sun Tzu to the present.[4] For the next two years or so I felt that I was too emotionally involved to write about him objectively, but then gradually came to the conclusion that I *could* do so and that knowing him well would be an asset in achieving a balanced view. Whether or not this would have proved the case in a biography I do not know because it was decided that it was too soon for this to be undertaken. Instead I was given permission to write the first full-length appraisal of his military thought, enjoying unrestricted access to his Papers, but with the stipulation that I should include only as much biographical information as was absolutely necessary to provide the context for his ideas. This I have been able to carry out without feeling that an emotional pull was interfering with my scholarly judgement; indeed some readers may perhaps think that I have leant too far the other way to avoid the charge of friendly bias. I can only say that I have not consciously leant one way or the other but have striven to be fair and impartial: to what extent I have succeeded must be decided mainly by informed readers, that is, readers familiar with the archives as well as the publications.

It will be obvious to such readers that this study makes no claim to be either comprehensive or exhaustive. What it does do is blaze a trail through the vast, and in some important areas virtually unexplored, forest of the Liddell Hart Papers.[5] Liddell Hart is universally recognized as the prophet of mechanized warfare or *blitzkrieg* and the champion of the strategy of indirect approach. It is also a commonplace that he had a lifelong interest in reforming the British Army, and that he achieved a good deal of success as unofficial adviser to Hore-Belisha on the eve of the Second World War. But there is much more to him than that. He attached greater importance to his more general reflective thinking about war and peace; this aspect of his work looms much larger in his archives than in his publications,[6] on which it provides many valuable insights; and above all what may be termed his 'philosophy' seems to me to possess more enduring interest than any particular theories or reforms linked with his name. I have therefore deliberately devoted more space to the development and influence of these general ideas than, for example, to the tactical and operational writings, which were directed primarily, though not exclusively, towards the professional soldier.

To keep up my earlier metaphor, this study will perform a useful service if it opens up new areas of the forest to other scholars. It is still far too early for a firm judgement about Liddell Hart's status as a military theorist; nor is it clear that his influence has ceased to be important. There is scope for several scholarly monographs on, for example, the extent and precise nature of Liddell Hart's influence on the British and certain foreign armies; his influence in shaping public opinion about the First World War in general and Sir Douglas Haig in particular;[7] his role as a military correspondent for *The Daily Telegraph* and *The Times*; the 'partnership' with Hore-Belisha; and his achievement as a historian. Such specialized studies would be an enormous asset to Liddell Hart's eventual biographer.

The present study constitutes the first attempt to put Liddell Hart's military thought in proper perspective by tracing the origins and development of his principal ideas over his whole career. The second volume of his own *Memoirs* terminates just after the start of the Second World War, and a projected third volume was never completed.[8] Another important aim of this study is to demonstrate that Liddell Hart's contribution to the understanding of contemporary problems of war and peace did not cease abruptly in 1940 or 1945. On the contrary, he was among the most prescient of the early analysts of the likely nature of war and conflict in the nuclear age. Only after 1960 did he for all practical purposes abandon the struggle to keep abreast of current affairs as an expert in order to devote his time increasingly to historical studies.

Anyone who has worked in the archives at States House will surely agree that Liddell Hart hoarded too much paper, yet at the same time they must be grateful for his obsession with preserving every scrap of written evidence. Thus, for example, in this study I have been able to draw upon his schoolboy notebooks and letters to his mother in sketching his character and outlook before the First World War. Further letters and an unpublished account of the Somme campaign reveal a rather unexpectedly romantic and patriotic subaltern lavishing praise on the British High Command. With the ending of the war and his involvement in the revision of infantry tactics the evidence already becomes voluminous: drafts of articles, annotated proofs of the various manuals and pamphlets, correspondence with officers engaged on the same questions of infantry training and with editors of journals—all have been sedulously preserved. General correspondence, however, remains patchy through the 1920s, mainly because he lacked a regular secretary, but there are a few important exceptions, notably Colonel J. F. C. Fuller and an American reserve officer, Captain J. M. Scammell, who briefly studied military history at Oxford in the 1920s. Liddell Hart habitually used his correspondence with such intelligent but not entirely like-minded officers to try out, refine and clarify his own ideas. Consequently their files have been heavily drawn upon to illustrate the development of his thinking in the crucial period immediately after 1918.

Here there is no need to do more than sketch his main concerns in the broadest terms. Unlike the majority of his contemporaries who had survived the holocaust, Liddell Hart became profoundly dissatisfied with the British military system and especially with 'the establishment' of senior officers who dominated the War Office. Yet he did not, like a radical minority, reject the system to the extent of becoming anti-military or (except perhaps towards the end of his life) a pacifist. Instead he was among the handful who tried to think coolly about how the stalemate and slaughter had occurred and how it could be avoided in the future.

> War was hell, but mere wishing would not prevent its recurrence. Somebody had to consider how, if it did occur, it could be fought more cleanly, more decisively, and above all more intelligently.[9]

This Herculean task he tackled methodically in the 1920s, first devising more fluid and mobile infantry tactics, next adding the revolutionary new ingredients of tanks and aircraft, and then broadening his scope to promulgate a strategic doctrine of 'indirect approach'. These ideas, particularly those involving the close combination of tanks and aircraft to achieve deep penetration into the enemy's rear areas and the strategic bombing of his homeland, had a decidedly *offensive* ring about them. Though eager to see them adopted by the British Services, the comparatively peaceful international situation in the 1920s permitted Liddell Hart to develop his *blitzkrieg* doctrines in a somewhat detached and scientific way. This, he seemed to be saying, was how modern, mechanized war should be conducted irrespective of who was fighting whom. His protagonists could just as well have been *Redland* and *Blueland*. In the 1930s this 'neutral' atmosphere rapidly disappeared and it became increasingly clear that whoever initiated the new style of warfare, it would not be Britain. In the operational sphere Liddell Hart consequently began to stress the superiority of the *defensive*, while in that of strategy and policy he urged reversion to 'the British Way in Warfare' which would involve only a 'limited liability' towards the Continent.

By the end of the 1920s, then, Liddell Hart had formulated and become identified with the clutch of principles, formulas and plain prejudices which served him as criteria for the rest of his life. Held with absolute conviction, they were a tremendous asset in providing him with 'keys' (a favourite metaphor) to the 'lessons of history' and to the interpretation of each new problem as it arose. Moreover they lent an air of scientific proof and certainty to his polemical writing. Inevitably, since he attacked boldly and often simultaneously on various levels from platoon tactics to grand strategy, his arguments were sometimes confused or even downright self-contradictory. In my view this was most clearly the case in the 1930s when, perhaps paradoxically, his reputation reached a peak. In the two chapters devoted to this period the basic documentation is provided by Liddell Hart's own books and newspaper articles. However, his correspondence files, notes on talks and memoranda in these years provide such a cornucopia that the problem was what to leave out.

The years between 1939 and 1945 are probably the most mysterious as regards Liddell Hart's activities and beliefs; yet the documentation is ample and extremely interesting. Some idea of his highly unorthodox views at that period can be gathered from the war commentaries which he threw together in volume form between 1940 and 1943. But these make for somewhat turgid reading compared with the numerous memoranda which he drew up and circulated to a wide range of people including ministers such as Eden, Halifax and Beaverbrook, an ex-Prime Minister in Lloyd George, Churchmen such as Bishop Bell, and a varied circle of personal friends. Here, as at other periods of his life, the more interesting revelations often occur in correspondence with comparatively unimportant individuals: for example, John Brophy, a writer serving in the Home Guard, and Esmé Wingfield-Stratford, a retired history don, both raised points which caused Liddell Hart to expound and defend some of his fundamental assumptions. By contrast more famous people tended to send formal acknowledgements and often did not even trouble to do that.

The essential feature of Liddell Hart's outlook during the Second World War was its consistency: he remained a root-and-branch opponent of total war and an advocate of a compromise peace short of victory which, he believed, would prove illusory even if it could be achieved. Before Dunkirk, Liddell Hart's view was quite widely shared in the country, but thereafter support for it dwindled rapidly. Liddell Hart's Papers reveal him as a virulent critic of Churchill and virtually all he stood for, so that he clearly condemned himself to remain in the wilderness as long as the latter was Premier.

In the 1940s international relations changed so rapidly that old heresies became new orthodoxies in the space of a few years, with the result that by the end of the decade Liddell Hart was in a strong position to say 'I told you so'. He had realized, soon after Operation Barbarossa failed in 1941, that a total defeat of Nazi Germany would result in Soviet hegemony in Central Europe with the added temptation to expand westward against comparatively weak defences. Whether the Western Allies liked it or not, Germany would have to be revived as a bulwark against Communism. In more general terms, his resolute opposition to total war became the conventional wisdom now that Hitler was defeated; while the advent of atomic weapons raised a real possibility of mutual annihilation to no purpose whatever. He had only to reiterate his philosophy of limited war, formulated in quite different circumstances after 1918, to put himself once again in the vanguard of progressive thought. He was among the first to realize that nuclear weapons, so far from abolishing conventional wars, might actually cause them to proliferate in non-nuclear areas of the world or as a result of the nuclear powers operating indirectly or by proxy. Even so, a new era in international conflict was beginning in which Liddell Hart did not feel intellectually at ease. On the one hand he was not deeply interested in the complex theory of deterrence which evolved to accommodate the startling developments in military technology in the 1950s; and on the other he remained utterly antipathetic to the rapid spread of guerrilla and revolutionary warfare, which he felt gave too much encouragement to indiscriminate violence and was always harmful to civil order in the long run, no matter how spectacularly successful in purely military terms.

It is curious that Liddell Hart's humanitarian feelings were seldom explicitly expressed in his writings. He was an extremely highly-strung and sensitive person, not, on his own admission, very brave in confronting pain in himself, and easily upset by illness and suffering in others. He felt a profound antipathy to violence in any form, and spent a good deal of his life in striving to ensure that if wars could not be avoided altogether they should at least be fought as humanely as possible. Yet in his published work the cool, dispassionate—almost clinical—tone is so consistently maintained that the horror and destructiveness of all warfare, however limited and skilfully conducted, is not brought home to the reader.

This cool, impersonal tone also carried over into public lectures and formal interviews. An Israeli journalist was clearly puzzled by the paradox of the military expert whose humane instincts made him utterly opposed to war:

> When you talk to Liddell Hart, you tend to forget that war means, first and foremost, blood, tears and sweat, and that its accompaniment is the groans of the wounded, the weeping of the widows, shattered bodies, destroyed buildings, bombed towns, great fear and pain, and whole worlds ruined at the touch of a hand.
>
> The Captain speaks of battles and wars and deadly weapons with a sort of cold academic calm, as if he were giving a lesson in Latin. It seemed to me at times as if he saw every war as a sort of game of chess, played on a world scale, according to set rules, in which victory goes to the cleverer, more cunning and stronger participant, as in the English proverb—let the better man win![10]

An explanation for this inhibition, if such it was, can only be speculative. The experience of the carnage in the battle of the Somme may have had such traumatic effects at the time that he did not care to dwell on it—though he was perfectly willing to talk about it later in life when I knew him. Alternatively, he may have made a conscious effort to set unpleasant memories and feelings aside in order to achieve objectivity as a tactical writer. A third, and to me more plausible, explanation is that he simply did not consider the overt expression of such emotions as horror, pity or compassion—over-indulged in the 'anti-war' literature after 1918—was appropriate in the professional writings of a regular officer. If this was the case, he remained in this respect more in sympathy with the outlook of the professional soldier than with the civilian critic of war *per se*.

Liddell Hart's Papers in his later years faithfully reflect his increasing preoccupation with military history, centred on the accumulation of evidence for his mammoth study of the Second World War. The documentation remains as bulky as ever but there is less and less general reflection in notes and memoranda, and the correspondence—though also vast—becomes more stereotyped.[11] On great events, such as the Arab–Israeli wars, he would circulate identical letters to dozens of friends and professional contacts. By the same token, the correspondence files shed very little additional light on his publications compared with earlier years. Thus the student who wishes to discover Liddell Hart's views on current problems *c.* 1950–1960 should go straight to *Defence of the West* (1950) and *Deterrent or Defence* (1960). I have not pursued my study beyond the latter date, but have concluded with an examination of his influence on the German and Israeli armies.

Notes

1. 'Liddell Hart: the Captain who taught Generals' in *The Listener* 28 December 1972. Michael Howard 'Liddell Hart' in *Encounter* April 1970.
2. See, for example, the article published under the transparent pseudonym of 'Bardell': 'Study and Reflection *v* Practical Experience' in *The Army Quarterly* July 1923.
3. Brian Bond (ed.) *Chief of Staff: the Diaries of Lieutenant-General Sir Henry Pownall, 1933–1944* (2 vols, Leo Cooper 1972–73: Shoe String Press 1973–74).
4. This assignment was subsequently taken on by Adrian Liddell Hart and published as *The Sword and the Pen* (Crowell 1976: Cassell 1978).
5. Exploration has been greatly simplified for future researchers by the admirable catalogue prepared by Mr Stephen Brooks and will be further assisted by the index currently being compiled by Miss Barbara Cave. For a good introduction to the history and present arrangement of the archives see Stephen Brooks 'Liddell Hart and his Papers' in Brian Bond and Ian Roy (eds) *War and Society Yearbook* II (Croom Helm: Holmes & Meier 1977) pp. 129–140.

6. He went through a pronounced 'philosophic phase' in the 1930s, possibly in part due to T. E. Lawrence's influence. Many of his reflections from this period are included in *Thoughts on War* (1944), one of his most self-revealing and interesting books.

7. Important insights on this subject are discussed by Paul Fussell *The Great War and Modern Memory* (O.U.P. 1975); see also Michael Howard's penetrating review in the *Times Literary Supplement* 5 December 1975.

8. Jay Luvaas' pioneering essay in *The Education of an Army* (U. of Chicago 1964: Cassell 1965) terminates in 1940, as does Irving M. Gibson's chapter 'Maginot and Liddell Hart' in E. M. Earle (ed.) *Makers of Modern Strategy* (Princeton U.P. 1941: O.U.P. 1944; republished unrevised as an Atheneum paperback 1966). See also bibliographical note, p. 279.

9. Michael Howard in *Encounter* op. cit.

10. 13/80 *Ma'ariv* 25 March 1960. Interview (in Israel) with Raphael Bashan. Liddell Hart marked the extract quoted but did not indicate disagreement with it.

11. Whereas, for example, up to 1959 there is at least one file per year containing memoranda, notes on talks etc., the years 1960–69 inclusive are covered by one thin file.

1

The Formative Experience, 1914–24

Basil Hart[1] was born in Paris on 31 October 1895 and did not live permanently in England until after the turn of the century. His father was a Wesleyan minister whose first appointment at Boulogne was followed by fourteen years caring for the spiritual needs of the Anglo-American and French Protestant communities in Paris. His mother's side of the family, the Liddells, came originally from Liddesdale on the Scottish border but had long been settled in Cornwall, where Basil's grandfather was closely associated with the development of the London and South-Western Railway, ending his career as Assistant General Manager. The Hart side of the family were yeoman farmers deeply rooted in the area of trans-Severn Gloucestershire and Herefordshire.

In 1903 the Harts settled at Guildford and Basil attended preparatory school, first at Edgeborough and later, at the age of eleven, transferred to Willington near Putney, where he was a contemporary of Maurice Bowra. He then followed his cousin Trevor at St Paul's. Basil's childhood and early youth were dogged by a series of illnesses and he was never to enjoy robust good health. In some 'Notes for an Autobiography' written as early as 1920–21, he remarked that his mother had nursed him through a whole group of childish complaints and 'rather spoilt me as regards hardening me. In fact she made me altogether too soft.' He was thus unprepared for the rigours of the typical Edwardian preparatory school and at Edgeborough was badly scared by the brutality of a gouty and vile-tempered headmaster.[2] In his early teens he became very tall and thin, outgrowing his strength, and this, coupled with illness, frustrated an early ambition to enter the Navy—he failed the medical test at the age of thirteen.

In his *Memoirs* he was to admit that he was mystified by the early development of his 'military bent' which first manifested itself in a precocious tactical interest in boyhood games. In addition to the games described in his *Memoirs* there survives an interesting essay in military fiction written when he was about fifteen and entitled *The Aquilon or the Conquest of the World by Aquilo.*[3] Set in the late classical period, the story provides early evidence of Basil's belief that history is made by great men, but in contrast to his later philosophy of war, Aquilo's victories were won at an enormous cost in casualties.

From his boyhood onwards Basil was deeply interested in a wide variety of games. At school he played football and cricket, though on his own admission without much natural aptitude; he did rather better at tennis, particularly at Cambridge; and during the First World War he discovered an unsuspected talent for long-distance running. In later life he played chess, table tennis and croquet with great dedication and amusement. These sporting interests are relevant to Basil's later career because they reveal his perennial preoccupation with tactics, and his tendency to criticize performance while at the same time extolling the exploits of the individual or star. These traits were later displayed in his journalistic covering of rugby matches and tennis tournaments.[4] A fascinating notebook survives from about 1910 in which Basil describes in great detail the imaginary history of Canadian rugby from 1914 to 1918 (why Canadian is a mystery). This saga is dominated by his hero Roland Havering Loring (a three-quarter like his real English hero of the pre-1914 era, R. W. Poulton) who in 1916 'stood head and shoulders above any other player . . . and won the game for Canada on his own'.[5]

But, like many schoolboys of that era, Basil Hart's predominant hobby in the period 1910–14 was aviation, whose military application was then of course only dimly foreseen. Here his imaginative approach again found expression in several notebooks filled with fictitious aviation competitions such as the Great 'Omnium' Meeting starting in October 1915 in which the leading 'aces' were Meredith (Britain), Brissac and Varvier (France), Ciro (Italy), Carteret (United States) and Kobo (Japan). In an aviation story written in the summer of 1913 he has inserted a later note claiming that he had accurately forecast a number of developments of the First World War, including all-metal and gun-carrying aircraft, and armoured cockpits to protect the pilots. Not content with fiction, he also conducted a regular correspondence with the editor of the magazine *The Aeroplane*, C. G. Grey, who, apparently not suspecting his age, published several of his letters and commented on them in leading articles.[6]

It is not intended to suggest that Liddell Hart's military genius was already evident in these early pastimes and writings: many schoolboys then, as now, presumably filled exercise books with notes on such topics as railways, aircraft, ships, cricket, football and birds. The notebooks do however provide early evidence of the intensity of his interests, his tendency to commit ideas to paper, his critical approach, his concern to predict future developments and—not least important—his self-confidence in believing that his ideas were worth publishing. In combination these interests and attitudes surely were unusual.

In the autumn of 1913, shortly before his eighteenth birthday, Basil Hart went up to Corpus Christi College, Cambridge, to read modern history. He did not work very seriously at the compulsory subjects but instead devoted a great deal of time to studying military history and working out tactical problems. He also continued his schoolboy hobby of writing to magazines and newspapers about aeroplanes and sport and had letters published in *The Morning Post*, *Evening Standard*, *Daily Telegraph* and *Football Evening News*. As a consequence of these distractions he obtained only a pass with third-class marks in the May 1914 preliminary examination.[7]

An autobiographical note in his pocket diary under January 1914 provides an early example of repeated attempts at rigorously objective self-analysis:

> Logical, self-love [crossed out] egotism, affectionate but not demonstrative, large brain-power, tactful and diplomatic, conventional, certain amt [amount] of individuality of thought, rather too methodical or even fussy, inclined to be philosophical not practical, head and heart fairly evenly balanced, too much love of detail—may fail to grasp the whole . . .[8]

Shortly after the outbreak of the First World War Basil Hart was among the thousands of young men who answered Kitchener's call for volunteers. Although the Sandhurst course leading to a regular commission had been reduced to three months, that was still too long for Basil, who shared the popular view that the war would be over by Christmas. He therefore took a temporary commission and began to train with the University O.T.C. He had joined the Army in defiance of his parents' opposition on the grounds that his health was too precarious. As his tutor Will Spens (later Master of Corpus) realized, he had been something of a spoilt mother's boy whom the Army 'would be the making of both physically and in character'. Basil was soon reassuring his parents that although O.T.C. training was rough and demanding he was revelling in it.[9] Indeed his first experience of a soldier's life brought out a curious military streak, in a 'credo' dated 28 November 1914 which would cause him wry amusement in later years. It was entirely in keeping with his respect for 'the record' that he should have preserved such an unflattering document which deserves to be quoted in full:

> Before the war I, Basil Hart, was a Socialist, a Pacifist, an anti-conscriptionist and an anti-disciplinist, disapproving of all state checks on the liberty of the individual and one who hoped for internationalization. I held thinkers in greater admiration than warriors.
>
> Now having studied the principles of warfare and undergone military training and seen the effects of it on my companions the following are my opinions:
>
> 1. *I believe* (i) in the supremacy of the aristocracy of race (and birth) (ii) in the supremacy of the individual.
> 2. In compulsory military service because it is the only possible life for a *man* and brings out all the finest qualities of manhood.
> 3. I have acquired rather a contempt for mere thinkers and men of books who have not come to a full realisation of what true manhood means. Military service if intelligently conducted develops and requires the finest mental, moral and physical qualities.
> 4. I exalt the great general into the highest position in the roll of great men and consider it requires higher mental qualities than any other line of life.
> 5. I consider the Slavs, by which I indicate a greater Russia, will rule both Europe and Asia and will have world dominion, being the finest and most virile civilisation and having the finest qualities of all races, and that the day of conquest and expansion is not yet over.
> 6. Socialism and its forms are an impossibility unless human nature radically alters.

7. There should be compulsory military service in order that all men may have the chance, which otherwise they would probably avoid, of developing true manhood.
8. Many of the German militarist ideas are very sound, but I oppose the Germans because I do not consider that the German type of mind is the one to carry out their ideas.

   I prefer brilliance to mechanical and methodical mediocrity, and that I do not consider the Germans have; and I consider that the Russians are of all nations the most likely to possess both brilliance and thoroughness. I certainly believe that absolute peace is detrimental to true manhood, but 20th Century war is too frightful. If you could have war without its *explosive* horrors it would be a good thing. I worship brilliance and brilliance seems to find its truest and fullest expression in the art of generalship.

   My belief in the necessary inferiority of women is more profound than ever.

   If the war ends by Easter it will be a great thing for the virility and manhood of Europe. If it continues till Xmas 1915 it will be a disaster.[10]

Like so many other fortunate survivors of the 'lost generation', Basil Hart was profoundly affected for the rest of his life by his experience on the Western Front. Yet in his case front-line experience was comparatively brief, being confined to two short spells in the autumn and winter of 1915 and a third on the Somme in July 1916. In his *Memoirs* he admits that his first spell near the front line, just north of the Somme sector, was 'a gentle introduction to war', and in a letter to his parents dated 29 September 1915 he remarked 'it reminds one most of a great picnic'. In mid-October he was struck by a sudden fever and had difficulty in persuading his doctors not to send him home. His second tour of duty later that winter introduced him to the Flanders mud of the Ypres salient and banished all enjoyment. His stay in the salient ended prematurely when he was concussed by a shell exploding above the doorway of a shallow dug-out in which he was sheltering. This time he raised no objection when the doctors marked him for immediate return to England.[11]

He returned to France in the spring of 1916 in time for the ill-fated Somme offensive, in which he took part from its opening on 1 July (though he was then in reserve) until 18 July, when he was badly gassed and 'knocked out' as his company was going out of the line. This brief but intense experience was later to provide an illustration of the shortcomings of British military leadership and tactics, but the complex psychological process by which he arrived at his critical outlook is difficult to document. Even though left out of the initial assault, he was hit three times without serious injury. His battalion, 9th King's Own Yorkshire Light Infantry, part of the 21st Division, was practically wiped out on the first day, which cost the Army nearly 60,000 casualties—the heaviest day's loss in British history. His battalion had only two officers left, one of them wounded, while a neighbouring battalion had none at all: Hart, though only a subaltern, took temporary command of the latter.

He was struck by the contrast between the notion of battle portrayed by war artists in the illustrated Press and the reality. As he recalled in his *Memoirs*:

> Instead of the dramatic charge of cheering troops which they depicted, one saw thin chains of khaki-clad dots plodding slowly forward, and becoming thinner under a hail of fire until they looked merely a few specks on the landscape.

He was also impressed at the time by the more sophisticated operational plan and tactics of General Rawlinson's Fourth Army on 14 July, consisting of a night advance followed by an assault at dawn preceded by a hurricane bombardment lasting only a few minutes. This plan, which secured surprise, ensured that 'grey-clad corpses outnumbered khaki ones on the battlefield'. 'That sight, and contrast,' Hart wrote in his *Memoirs*, 'deeply influenced my future military thinking.' At the time, however, he was completely uncritical and regarded the battle as a victory despite the heavy losses.[12]

It is important to establish that during the period of his actual service Basil Hart was very much the conventional, ultra-patriotic ex-public-school officer whose admiration for the High Command bordered on adulation. Thus in a 'last letter if killed on duty' to his parents dated 27 May 1916 he requested that in any memorial he would like it to be stressed that he was going to the front for a third time. He added that while it was an honour to die for England, 'I feel it an even greater honour to die as an officer of the British Regular Army.' Shortly before the Somme battle began he described Haig and Robertson as 'our finest leaders' and even in July 1916 he described British military organization as 'brilliant', while generalship and staff work were 'marvellously clever'. His undiscriminating list of 'great British generals of the war' reveals that his schoolboy tendency to hero-worship was still very much alive.[13]

This hero-worshipping phase reached its apogee in a little book he wrote during convalescence in September 1916 entitled *Impressions of the Great British Offensive on the Somme* by 'A Company Commander who saw three and a half weeks of it'. There could hardly be a more comprehensive catalogue of the assumptions about British generalship in the First World War which he would denounce with increasing severity for the remainder of his career.

Thus, to quote just a few examples, he praised the 'amazing perfection of our organisation which in its generalship and staff work were super-German'. The Somme area was well-chosen for the offensive and the strategy was masterly. Sir Douglas Haig was 'the greatest general Britain had ever owned'. 'Whilst not all our staff are brilliant, it is safe to say that 90 per cent of our general staff officers are really brilliant men, with quite a large number of men amongst them who have a genius for war . . . We have produced fully a hundred first-rate generals.' He denounced ignorant civilian critics of the war who only revealed their shallowness of mind: 'it is only the General Staff who really understand the value of any operation'. More generally he concluded that:

> War, at least modern war, as waged on the Western Front, is horrible and ghastly beyond all imagination of the civilian. Nevertheless it has an awe-inspiring grandeur of its own, and it ennobles and brings out the highest in a man's character such as no other thing could. Could one but remove the horrible suffering and mutilation it would be the finest purifier of nations ever known. Even as it is, it is the finest forge of character and manliness ever invented, when taken in small doses. The unfortunate thing is, that this war has become an over-dose. Still, with all its faults and horrors, it is above all a man's life, in the fullest and deepest sense of the term.[14]

Perhaps it would not be too unfair to say that he never

entirely abandoned the sentiment of this concluding passage, but strove to ensure that in future war could once more be taken 'in small doses'.

Although this patriotic contribution to the war effort was never published—the War Office rather surprisingly refused permission—it was far from being a wasted effort. The author sent a copy to his supervisor at Cambridge, Geoffrey Butler, who in turn showed it to a number of friends, including John Buchan, who was then on the staff at Haig's headquarters. Buchan, himself a fervent admirer of the British High Command, was impressed by the typescript and invited Hart to join the historical section which he planned to set up at G.H.Q. This scheme—like so many in Liddell Hart's early career—failed to materialize because of Buchan's illness and subsequent transfer to a different appointment at home, but Buchan later proved a very useful contact. From the very start, in fact, Liddell Hart displayed a flair for employing his writings to make contacts and friends in influential positions. For example, during the last two years of the war, when serving as Adjutant to Volunteer units in Stroud and Cambridge, Captain Hart (as he was then) wrote several booklets on infantry drill and training which he circulated to good effect to impress several well-placed officers, including his first and perhaps most important patron, General Sir Ivor Maxse, who had been Inspector-General of Training in France during the last six months of the war and was then appointed General Officer Commanding-in-Chief, Northern Command. Not only was General Maxse instrumental in giving Hart's ideas on infantry tactics and training wide publicity in the Army, and finding him employment in his command; but he also opened up an outlet into serious journalism through his brother Leo Maxse, who was editor of the monthly *National Review*. It was in this magazine, at Sir Ivor Maxse's suggestion, that Captain Hart's important 'Man in the Dark' theory of tactics was published in an expanded version in June and July 1920.[15]

It is not the purpose of this study to chronicle Liddell Hart's career in detail, since that has already been admirably done in his *Memoirs*. Suffice it to remark here that by 1920 Liddell Hart had already begun to make a name for himself, first as a fine trainer of amateur soldiers and then as an expert on infantry tactics, particularly at the section, platoon and company levels.[16] Yet all the time his survival in the steadily dwindling regular Army was precarious because of his impaired health. On this front he fought a valiant but vain struggle until 1924, when he was invalided out of the service on half pay.

As hinted at earlier, Liddell Hart's uncritical eulogy of British generals and generalship was but a transient phase; not surprisingly, in his *Memoirs* he confesses he was amazed at the enthusiastic praise he had bestowed on Haig and others. Replying to a later critic, he argued that this unpublished work proved that he had no bias against the Higher Command, but rather a strong inclination to credit it with infallibility until eventually disillusioned by post-war study of its plans and operations. Elsewhere in the *Memoirs* he pointed out that, contrary to what many people later imagined, he was not an instinctive rebel.[17]

Nevertheless it would be more accurate to speak of a revival and sharpening of his critical instinct, for its spirit was certainly not new and had only been temporarily stifled by romantic patriotism. Probably he had inherited a strong sense of curiosity, high moral standards and a desire to analyse; while his tendency to question orthodox views had certainly been encouraged by an eccentric but stimulating master at St Paul's, the legendary Reverend Horace Elam.[18]

In trying to trace the emergence of Liddell Hart as a trenchant and formidable military critic one can appreciate how a rigorous re-examination of infantry tactics would lead him to revise his original eulogies of generalship and staff work. Yet a greater impetus seems to have come from another direction. This was summed up in a note he wrote in 1920 or 1921 on 'How my hero-worship of "generals" waned and disillusionment began, through close contact with the *best* of them—and finding their lack of fresh ideas—how they depended on a novice like me to show them the lessons of the war'. The *Memoirs* contain numerous examples of this disillusioning process which strengthened Liddell Hart's high estimation of his own abilities. In short, his wartime heroes were found to possess not merely feet of clay but wooden heads as well. Thus Lord Horne, whom he had imagined to be a superman on the Somme, was found to be very obtuse when Basil tried to explain to him the use of the sand table in infantry instruction when he visited Shorncliffe in

1922. Worse still, he later opposed Liddell Hart's promotion with the comment that 'writing on military subjects does not justify accelerated promotion'. A few years later General Montgomery-Massingberd epitomized Liddell Hart's notion of a reactionary commander when he first abused a new book by Colonel J. F. C. Fuller and then admitted he had not read it, saying: 'No. I have not read Fuller's book! And don't expect I ever shall. It would only annoy me!'[19]

If personal experience impelled Liddell Hart towards a critical stance, his critical attitude was strengthened both by his own intensive studies of the First World War beginning in the 1920s, and by the publication of numerous diaries and letters (such as those of Field Marshal Sir Henry Wilson) which revealed the in-fighting and disloyalty that occurred amongst the senior commanders. Liddell Hart was clearly appalled and disillusioned by the commanders' *moral* shortcomings as well as what he regarded as their professional incompetence.

In his rapid transformation from eulogist to waspish critic, Liddell Hart to some extent lost his sense of historical perspective. He gave insufficient weight to the long-term effects of the tradition of the strict subordination of the British Army to political control since the seventeenth century, and to the distinctive character of the British officer corps. Since the great majority of first commissions in the infantry and cavalry and promotions up to the rank of lieutenant-colonel were subject to purchase until 1871, and even thereafter a private income of £200 and upwards per annum was still essential for officers in many regiments, it is easy to understand why in some respects an amateur spirit persisted. For a considerable number of regimental officers before the First World War, soldiering was a temporary occupation for the gentleman rather than a life-long career. This type of temporary officer tended to despise such professional hurdles as the Staff College which had to be taken by the ambitious. Also at times Liddell Hart perhaps overvalued the intellectual, critical outlook in officers—or rather expected too much of the average officer—at the expense of other qualities of character.

Even so, on balance—and it is a very difficult balance to strike—Liddell Hart's exceedingly critical view of the British officer corps between the two World Wars seems to the present writer to be justified. One difficulty is that brilliant, unorthodox and outspoken critics of the military establishment such as Liddell Hart (and even more Fuller) tended to provoke the 'Blimpish' reactions like that of Montgomery-Massingberd quoted above and thereby confirmed their rather stereotyped notion of reactionary generals. On the other hand a host of more progressive regular officers privately endorsed Liddell Hart's criticisms of some of their seniors. When every allowance is made for the element of self-justification, Liddell Hart's *Memoirs* and archives remain an impressive indictment of the mentality of the 1914–18 generation of senior Army officers.[20]

In later life Liddell Hart devoted endless time and energy to encouraging promising young men to persevere along their chosen course no matter how many and depressing the setbacks. These kindly words and acts were doubtless in part inspired by his own experience in the half-dozen or so years after the First World War when his health was poor and his luck appeared to be out.[21] Although in retrospect those years may be seen to mark a steady and rapid—even meteoric—rise to distinction as a tactical theorist, he in fact suffered numerous setbacks and disappointments which probably helped to strengthen his conception of himself as a genius frustrated by petty-minded individuals and bureaucratic red tape. In 1921, for example, he discovered in going through a War Office file about the Infantry Training Manual that an officer with whom he had co-operated closely in revising the manual had written 'a cool and cold-blooded note' to a respected senior officer, Brigadier Winston Dugan, suggesting that they should 'pick Liddell Hart's brains but need not trouble to do anything for him in return'! It is greatly to Liddell Hart's credit that he seems to have harboured no resentment against the author of the note, Captain (later Lieutenant-General Sir John) Evetts.

> After the war [as Jay Luvaas concisely puts it] Liddell Hart served in various capacities as education officer organizing educational work in hospitals, helping to rehabilitate disabled veterans in civil life, and teaching military history to officers studying for their promotion examinations. He sought an appointment to the historical section, Committee of Imperial Defence, but was turned down because the economy-minded financial branch would not allow an officer to go to a government department on pay from army funds.

> He applied for a commission in the newly formed Army Educational Corps but was rejected by the medical board, and when friends in high quarters intervened to have him declared physically fit, the best the selection board would offer was a regular commission as lieutenant.

In August 1923 he was notified of his selection for transfer to the Tank Corps in the face of keen competition, but unfortunately once again was not passed medically fit.[22] At that time he contemplated looking for work in the United States as a lecturer and freelance writer.[23] His luck at last took a turn for the better in 1924, as will be related at the end of this chapter.

Liddell Hart's professional reputation was founded on his radical reformulation of infantry tactics which began towards the end of the First World War and continued into the early 1920s. Here he was building on solid foundations closely related to his own experience. His own account of this process in his *Memoirs* deserves to be quoted at length:

> In the development of my military ideas the experience of having to train a battalion from scratch played a greater part, as I have come to realise, than such experience as I had acquired on active service in France. For it was a very different problem, and a much harder one, than to carry on the training of troops habituated to the practice and procedure of warfare on the entrenched Western Front . . .
>
> The need was to develop a tactical training system for attack, defence, and counter-attack, to fit mobile conditions in a way that would reconcile the fundamental principles drawn from past experience with the recent development in weapons. So I was impelled to think out . . . the reasons for the tactical precepts and practices which had been taught in 1914 and the following years. The more I thought about them, the more I came to doubt their validity in numerous respects, and to feel the need for a more systematic exploration of military theory as a whole, in relation to military history. So I started from the bottom, the field of minor tactics familiar to me, working upwards to the theory of *la grande guerre* . . . and then downwards again to the level of minor tactics.[24]

His quest for the utmost theoretical simplicity led him to postulate the situation of an unarmed 'man in the dark' confronted by a single unarmed opponent. He deduced that success, at this and higher levels, depended on a compound of fixing, manoeuvre and exploitation. These ideas were developed in an important article in the *United Service Magazine* in April 1920. The essence of his theory was that:

> In war we resemble a man endeavouring to seek an enemy in the dark, and the principles which govern our action will be similar to those which he would naturally adopt. The man stretches out one arm in order to grope for his enemy (Discover). On touching his adversary he feels his way to the latter's throat (Reconnoitre). As soon as he has reached it, he seizes him by the collar or throat so that his antagonist cannot wriggle away or strike back at him effectively (Fix). Then with his other fist he strikes his enemy, who is unable to avoid the blow, a decisive knock-out blow (Decisive Attack). Before his enemy can recover, he follows up his advantage by taking steps to render him finally powerless (Exploit).

The main idea he strove to inculcate was that 'weight or force' in modern war ought to mean weight of fire power and not of mere numbers of men:

> While manoeuvre is the key to victory, it is manoeuvre of the units of fire power and not of masses of cannon fodder. We must learn to depend for success, not on the physical weight of the infantry attack, but on the skilful offensive use in combination of all available weapons, based on the principle of manoeuvre.

In future, instead of masses of men with bayonets—

> one will send to the attack widely dispersed combat groups, containing comparatively few men but amply equipped with fire power, supported, moreover, by masses of auxiliary firepower, such as artillery, machine guns, tanks, and land fighting aeroplanes.[25]

An expanded version of the 'man in the dark theory of war' was published in the *National Review* in June and July 1920. In the first instalment he stressed the potential value of the caterpillar track used not only on tanks but on all forms of transport in the battle zone to revive the possibility of exploiting tactical success. If this could be achieved it would again be possible to garner the full fruits of victory as Napoleon had done at his best. Henceforth, when caterpillar transport became universally adopted, armies would be able to go straight across country and there would no longer be rigid lines of communication for the enemy to block by bombardment from guns or aeroplanes. Warfare, in short, would become more mobile than ever before.[26]

When Liddell Hart was given the opportunity to prepare a new edition of the *Infantry Training Manual* in the summer of 1920 he could draw upon the improved tactical methods developed in 1918, first by the Germans in their March offensive and then by the Allies.

The German offensive of March 1918 had demonstrated that even without tanks the defences could be broken by specially trained infantry closely supported by artillery. On Ludendorff's orders the German divisions transferred from the Eastern Front had undergone intensive offensive training in accordance with Captain Geyer's new handbook *The Attack in Trench Warfare*. The youngest, fittest and most experienced soldiers were formed into élite units and known as Storm Troops. Armed with light machine-guns, light trench mortars and flame-throwers, their task was to employ infiltration tactics to cross the trench-lines, bypass centres of strong resistance, and if possible, penetrate so deeply that they could attack the enemy's artillery. Contrary to the customary practice on the Western Front, reserves were to be thrown into battle where the attack was progressing, not where it was held up.

Ludendorff's tactics, supported by a gigantic artillery barrage, gained impressive though not decisive successes against British troops who were sadly unprepared for open warfare. What is not generally appreciated, however, is the speed and efficiency with which the Allies learned their lesson. This was chiefly due to Haig's newly appointed Inspector-General of Training, General Sir Ivor Maxse, 'a man of imagination, originality and drive with a wealth of battlefield experience behind him'. The new spirit instilled by Maxse and his staff in the Army and Divisional Schools was made manifest as early as 4 July by units of Rawlinson's Fourth Army, Monash's Australian Corps and the United States 33rd Division in the battle of Hamel. Success on a much larger scale was achieved in the battle of Amiens on 8th August. Although tanks played an important part in this victory, Fuller and Liddell Hart perhaps exaggerated their role at the expense of the artillery and infantry. As General Hubert Essame justly summed up: 'Victory in fact had been the reward of all arms skilfully acting in combination with one simple aim'. The same author has described the smashing of the Hindenburg Line by Haig's Army at the end of September as 'a gigantic operation surpassing in size anything before 1945 in World War Two'.[27]

These tactical lessons may seem very clear to historians in retrospect but Liddell Hart fully deserved the credit, which he claimed for himself in his *Memoirs*, for first grasping the essential principles behind the operational experience of 1918 and then developing them, in terms of his own original concept of 'the expanding torrent', for training purposes. There is no suggestion in General Sir Ivor Maxse's correspondence that he believed Liddell Hart to be merely codifying his own successful wartime tactics: on the contrary there is ample evidence that it was the twenty-four-year-old subaltern's penetration and originality which won the General's admiration.[28]

Liddell Hart explained the need for his 'expanding torrent' system of attack in a letter to Captain Evetts, Maxse's G.S.O.3., in August 1920:

> In thinking over what we had previously written and the old 'soft spot' idea, it seemed to me that the latter left off at the critical moment—just after the soft spot had been made and the reserves pushed towards it. It said nothing towards the vital problem of breaking through a system of defence in depth. A large body of reserves were to go through a bottleneck and no hint was given as to the subsequent action.[29]

In 1925 Liddell Hart argued persuasively that his 'expanding torrent' idea differed from the infiltration method of attack evolved by the Germans and adopted by the British in 1918 in three respects:

> 1. That while the reserves are pushed in along the line of least resistance they are used to expand the penetration and to maintain the original breadth and pace of the advance coincidently.
>
> 2. This dual purpose is fulfilled by an automatic backing up of those forward units which are making quicker progress, and by a progressive replacement of the slower forward units alongside them...

3. Movement by the line of least resistance is automatic, so saving time, but the superior commander is able to retain control of the rate and direction of the penetrative advance.[30]

In these early publications on infantry tactics Liddell Hart sought simplicity in the hope that his principles could be grasped by N.C.O.s and soldiers as well as officers and become so ingrained that they would be implemented automatically in battle. As General Maxse put it, in approving Liddell Hart's 'ten commandments' for infantry tactics in 1919: 'The above reply to your letter is easy. What is difficult, and will require ten years of strenuous endeavour, will be to get the doctrine understood, adopted and actually practiced over a scattered Empire like ours! It is a big *training* task—and implies a mental effort on the part of numbers of officers who dislike thinking. But, we can do much to *start* on good lines.'[31]

Liddell Hart stressed that the basic principles should also be applicable to any unit in all forms of action. The keynote he repeatedly struck was that henceforth tactics must be based upon 'the intelligent manoeuvre of fire power'. Dispersion, flexibility and individual initiative would be at a premium in the battles of the future. He was also in the vanguard of progress in thinking in terms of functions rather than the roles of the traditional arms. Although Liddell Hart frequently mentioned tanks in these writings—and indeed as early as November 1919 had published an article on infantry and tanks in a combined unit[32]—his main preoccupation was with infantry, which he believed was still potentially capable of exploiting a breach in an entrenched front. This was not all that surprising, given Liddell Hart's very limited personal experience and pride in his regiment—the King's Own Yorkshire Light Infantry. What is more surprising, considering its dominant role throughout most of the First World War, is his comparative neglect of artillery, not only in his immediate post-war writings but throughout the 1920s.[33]

Liddell Hart's complete conversion to the opinion that tanks would constitute the decisive weapon in future warfare stemmed mainly from his association with Colonel J. F. C. Fuller, which began in 1920. Fuller, while Chief Staff Officer at Tank Corps headquarters, had devised the revolutionary 'Plan 1919'. This envisaged a surprise attack by medium tanks on a ninety-mile front designed to penetrate twenty miles in order to paralyse the German divisional, corps and army headquarters. This broad-front break-in would be followed by a more concentrated thrust on a fifty-mile front exploited by a pursuit force of tanks and infantry carried in lorries. Fuller calculated that as many as five thousand heavy and medium tanks would be required for the whole operation—truly a vision of the distant future in 1918. He also planned for the R.A.F. to provide reconnaissance and tactical assistance. The war ended before this prototype of modern *blitzkrieg* could be implemented, but Fuller continued to extol the supreme importance of tanks in lectures, articles and books.[34]

Liddell Hart opened their long and mutually beneficial correspondence by sending Fuller copies of his two articles on 'The "Man in the Dark" Theory of War' (from the *National Review*) for comment. Apart from the criticism that Liddell Hart had made too much of the 'fog of war' by 'converting it into pitch darkness', Fuller stressed that tanks were likely to be the key weapon in pursuit, since cavalry was now practically useless while infantry was too slow. On the latter point, after an unusually deferential opening ('If I might dare to couple my humble self with you') Liddell Hart replied:

> I am writing for the immediate present, while infantry are still considered the chief arm. I feel quite certain that the tank will very soon become the sole arm of importance, if only the authorities will listen to the people with organised brains, which naturally include foresight, of whom you are the chief.[35]

Early in 1922 Liddell Hart was obliged to present the best arguments he could find against the mechanical warfare school when he agreed to draft the article on 'Infantry' which appeared under General Maxse's name in a new edition of the *Encyclopædia Britannica*, but he admitted privately that Fuller's criticisms had undermined his confidence in the case for infantry. 'Infantry is more likely to endure,' Liddell Hart wrote to Fuller, 'because of conservatism, financial and official, than its own inherent merits. In the meantime one must do one's best to improve the infantry's technique and raise its morale. Of course although tanks are unquestionably the weapon of the immediate future, man as an individual fighter may be resurrected. The infantryman's justification rests on two factors. He is the smallest target and the most universally mobile of all weapon carriers.' The letter concluded with a prophecy:

> The nation which is quickest to realise that the period on which we are entering is the tank era will win the next war.

On receiving Fuller's detailed criticisms of his article on 'Infantry' Liddell replied: 'I am an ardent believer in tanks and only want to be converted entirely. Full conviction is however necessary if one is to be a good advocate.' After carefully examining Fuller's points he admitted frankly that, 'I am nearly in agreement that my task, to uphold infantry against the inroads of mechanical warfare, is impossible.' Liddell Hart's only reservations concerned the use of gases, wireless interference and armour-piercing bullets against the tank. Otherwise 'Your arguments are so convincing on the tank *v* other arms as they exist, that I am fain to become a disciple. I was not at Oxford, "the home of lost causes". If it is not trespassing too far on the kind interest you have taken in my efforts may I ask what are the possibilities of a transfer to the Tank Corps?'[36]

Thus from 1922 on Liddell Hart wholeheartedly joined Fuller in spreading the gospel of mechanization in which the tank would become the chief criterion of military progress. There remained however an important difference in their conception of armoured warfare. Fuller tended to argue that the tank alone would dominate future battlefields and that the infantry's task would merely be to occupy the territory won by the tanks. Liddell Hart on the other hand maintained consistently that there was need for a more mobile type of infantry—what he called 'tank marines'—to co-operate with tanks in an armoured force for prompt aid in overcoming defended strong points. In short, Fuller concentrated on the development of an all-tank army, while Liddell Hart favoured an all-mechanized army in which all the supporting arms would be mounted in armoured vehicles, and would thus be able to accompany the tanks closely.[37]

Liddell Hart's deferential tone in his early correspondence

with Fuller has already been mentioned. Since the former was scornful of false modesty and did not suffer from it himself, his fulsome praise of Fuller must be regarded as entirely sincere. Indeed Fuller was not only the senior in age and rank; in the 1920s he was also the more dynamic and probably the more original thinker of the two. Fuller's voluminous correspondence with Liddell Hart, and his prolific publications at this period on both the history and the theory of war, reveal a range and fertility of ideas that would be remarkable in an officer of his experience in any army. Thus Liddell Hart was not exaggerating when he described Fuller to a keen American student of military affairs, Captain J. M. Scammell, as 'the greatest intellectual power I have ever come across, a triton amongst minnows'. When Liddell Hart began to draw up lists of promising British officers with 'star' ratings in 1925 Fuller was in a category by himself as the only one to score a maximum of three stars. When, in 1928, the two had a tiff, exacerbated by an angry exchange between their wives—neither of whom was noted for that—Liddell Hart wrote at length to put the record straight. Although pointing out, *pace* Sonia Fuller, that he owed his first steps up the ladder to Maxse rather than her husband, he generously conceded Fuller's pre-eminence:

> you were the pioneer [in mechanization] and ... my conversion did not begin until 1918 and was not complete till 1921. Up to that time I was essentially a pioneer in the field of infantry tactics and had hardly studied mechanical warfare. I have long considered that yours is the profoundest intellect which has been applied to military thought in this century. Although I feel my own mind is progressively developing in power and range, I have always recognised your superiority ...[38]

Unfortunately, even before his association with Sir Oswald Mosley's Fascists in the 1930s made him politically suspect, Fuller had sacrificed his chances of reaching the top of the Army by his arrogant behaviour towards his superiors in rank and by his reckless outspokenness. At times he almost seemed to cherish a death wish so far as his military career was concerned.[39] In the 1920s Fuller may be described as a blazing comet, while Liddell Hart was a less brilliant but steadier star.

Liddell Hart was sufficiently certain that he had already

achieved a niche in history to compile *Notes for an Autobiography* in 1920.[40] His motives for writing and preserving these and other self-assessments must remain a matter of speculation. The facts that his mother had made him 'too soft' in infancy, and that his health as a young man had been poor as a result of outgrowing his strength, may have provided extra incentives to achieve distinction in public life. On his own admission, as an undergraduate at Cambridge he was very diffident about his abilities and had quite a low opinion of himself. Later he welcomed the opportunity to give orders as an officer, not—as he wrote—that he was ever unusually shy at it, but was lacking in self-confidence. What is clear and worth establishing at this early point is that his craving for praise and recognition of his 'greatness' is a constant trait in his character and not something he developed later as a reaction to the temporary eclipse of his reputation during the Second World War.

In 1921 the main items in a note entitled *What I have achieved by the age of 25* were as follows:

1. I have lectured on tactics at the Royal United Services Institute and the Institution of the Royal Engineers...
2. I have written the whole tactical half of the Official British infantry manual.
3. I have written the article on infantry tactics for the new edition of the Encyclopaedia Britannica which, as the announcements say, 'is written by the greatest living authorities on each subject'.
4. I have invented a new method of attack, a new method of defence, a new system of battle control besides 70 lesser ideas on infantry tactics...[41]

Finally, Liddell Hart's self-analysis of December 1920 is so revealing that it deserves quoting at length:

> What audacity to try and comment on oneself! A very clear analysing brain. Not very retentive or persistent, but fond of getting down to the root of everything and then crystallising and simplifying it into a clear framework. Original in ideas, with no prejudices of any kind. Very sympathetic and keen to understand other people's point of view—to which I owe my ability to handle men. Very generous, in fact foolishly so.

Highly strung and sensitive, though I have learnt by experience to conceal it. Not physically brave. Hate to take blind risks and jumpy, but perfectly all right as soon as I have had a moment to seize control of my nerves. As my nerves are usually on edge, I have learnt the art of always being ready to control them, and successfully so far. Yet in the sudden emergencies which have befallen me—mostly in motoring—I have usually kept my head and instantly done the right thing, which has earned me a reputation for coolness.

Very good in argument, always being quick to seize the weak points in my opponent's case and turn them to success. Normally lazy, but when the mood or the need comes ready to use my last ounce. Model my life in all matters on the two tactical principles of Economy of Force and Security. Inclined to be selfish and think of myself first, and yet often going to the other extreme. Extremely broad-minded. A peaceful and philosophic temperament, but when really roused, become almost blind with rage, though I never completely let myself get out of hand. Generally thinking ahead of the immediate present. Always write on any subject, particularly military, with one eye on the historian of the future.

Like power, but to wield it for the good of those under me. Would prefer a small niche in the temple of historic fame, to an evanescent flame of present glory. Idealistic and romantic, but practical. Artistic, but no idea of art or music.

One thing I can put on record, and that is that in all my meetings and arguments with the pick of our Regular soldier brains, I have been able to out-argue and out-think them, fine as they were, and have been conscious that mentally my brain was both deeper, clearer, and more agile than those I have encountered...

I realise fully that opportunity and luck play a great part in a successful career, but I do feel that given health and opportunity I could at the present time handle a brigade at least with distinction in battle and that with experience I could prove myself one of the 'masters of war'. But under modern conditions it looks as if the opportunity, if it ever comes, will only come—under the rigid rules of seniority—when my intellectual keenness and agility is dulled by time. I feel quite certain that even at the present moment I could command an infantry brigade in war at least as well as any brigade commander at present, and I believe that with more experience of the conditions and capabilities of other arms I could command an army equally well if given the chance.[42]

The year 1924 marked a watershed in Liddell Hart's career. His hopes of remaining a professional soldier ended when he was invalided to the half-pay list—incidentally a misleading term since in fact he received barely a quarter of his pay. His health was, to say the least, uncertain and also he now had a wife and young son to support. He had no academic qualifications, and strenuous efforts by his former teachers to secure him an opportunity to obtain a doctorate at Cambridge on special terms not surprisingly failed because he could not afford to go into residence there. As on previous occasions, Sir Ivor Maxse proved a most valuable friend in need by securing for Liddell Hart the opportunity to cover the Territorial Army manoeuvres for the *Morning Post*. This task he performed so well that in 1925 he was chosen to succeed the famous Colonel Repington as military correspondent of *The Daily Telegraph* against keen competition, including several well-known generals. This appointment, which he held for ten years, gave him an excellent platform from which to press for the modernization of the Army in general and for the development of mechanized forces in particular. In addition it had the advantage of affording him a certain amount of leisure and personal freedom to pursue his broadening interests in the study of war.

By the mid-1920s Liddell Hart was also ending one phase of his intellectual development and entering another. He had virtually completed his theoretical analysis of infantry tactics and was now becoming preoccupied with the need for mechanization. Already in October 1924 he had outlined the case for wholesale mechanization as a long-term goal in his article 'The Development of a New Model Army'.[43] Similarly he had exhausted his exploration of the essential 'principles of war' based on the actions of the individual combatant—the 'Man in the Dark Theory of War'.

He was now becoming more concerned with war as a social and political phenomenon, and in the operational sphere with strategy rather than tactics. Growing dissatisfaction with the conduct of the First World War provided the chief impetus for these broadening studies. Already by 1924 he had fastened on the notion that Clausewitz was the evil genius whose false strategic doctrine was responsible for futile battles of attrition such as Verdun, the Somme and Passchendaele.[44] Simultaneously he was also devouring freshly available evidence about the First World War and exploring the wars of earlier centuries in quest of a general theory of strategy. This quest and its outcome provide the main theme in the next chapter.

Notes

1. He did not adopt the style 'Liddell Hart' until 1921. It is noteworthy that after leaving the Army he retained his modest rank for literary purposes and signed himself 'Captain B. H. Liddell Hart' until knighted in 1966—hence Yigal Allon's tribute to 'The Captain who teaches Generals'.

2. 'Notes for an Autobiography written in 1920–21' 7/1920/38. See also *Memoirs* I (1965) pp. 2–10. (All references are to this volume unless otherwise stated.)

3. 7/1910/1.

4. See for example *The Lawn Tennis Masters Unveiled* (Arrowsmith, London 1926).

5. 7/1910/1; Notes for Autobiography op. cit. p. 2.

6. 7/1913/5, 6, 12.

7. 8/28 Geoffrey Butler to Liddell Hart 18 June 1914. His marks were as follows: Essay 35, Constitutional History 33, Economic History 43, Mediaeval History 45. (Fail mark 33, Second Class mark 45.)

8. 7/1913/13.

9. 7/1914/2–4 Liddell Hart to his mother, n.d. but autumn 1914. He repeated to her Will Spens's remark that he was one of the six for whom Army training would be beneficial in terms both of physique and of character.

10. Note dated 28 November 1914, 12/1914/10.

11. 7/1915 *passim. Memoirs* pp. 13–18. See also Liddell Hart's essay 'Forced to Think' in G. Panichas (ed.) *Promise of Greatness: the War of 1914–1918* (John Day: Cassell 1968).

12. Compare his contemporary description of his experiences in the Somme battle in letters to his parents (7/1916/10–16) with *Memoirs* pp. 18–26. In a letter home dated 4 July 1916 he cheerfully reported that whereas before the war he could hardly bear the sight of a cut finger, now he could stand the sight of mangled limbs etc. (7/1916/11).

13. 7/1915/9, 7/1916/8, 11, 21, 34. Under the pseudonym 'Regular Officer' he published a column in the *Daily Express* extolling the 'Great Generals of the War' (7/1916/35).

14. 7/1916/22 pp. 1, 30, 35–38, 40, 43, 76, 84–85, 93.

15. 7/1920/87 Sir Ivor Maxse to Liddell Hart 9 April 1920. *Memoirs* pp. 26–27, 38ff., 47, 57. 'The "Man in the Dark" Theory of War' in the *National Review* June 1920 pp. 473–484 and July 1920 pp. 693–702. The author was paid £1 per page.

16. In addition to booklets on discipline and duties of a section commander in the attack, Liddell Hart published in 1918 *Battle Drill or Attack Formations Simplified* (11 pp.) and *New Methods in Infantry Training* (38 pp.). See also *The Framework of a Science of Infantry Tactics* (1921).

17. *Memoirs* pp. 26, 60. Nevertheless, if not an 'instinctive rebel', he later liked to stress his individualism and unorthodoxy—e.g. on a master's adverse report from his preparatory school in 1910 he has added: 'He frowned on my unorthodox style of bowling and goal-scoring!!' (8/19).

18. *Memoirs* p. 9. Elam is depicted in fictional form in Compton Mackenzie's *Sinister Street* and Ernest Raymond's *Mr Olim*.

19. *Memoirs*, pp. 58, 102.

20. It is interesting to compare Liddell Hart's views with those of Arthur Marder on the Royal Navy's senior officers at the same period: see Marder's *From the Dardanelles to Oran* (O.U.P. 1974) pp. 33–63. See also A. P. C. Bruce 'A History of the Purchase System in the British Army, 1660–1871' (unpublished Ph.D. thesis, University of Manchester 1974); and Brian Bond *The Victorian Army and the Staff College* (Eyre Methuen 1972).

21. Heart attacks in 1921 and 1922 left an obsession that he might suddenly drop dead (8/317). Although he complained of 'bad luck', from the evidence of his *Memoirs* it seems that the various medical boards which refused to pass him fit were only doing their duty. Fuller's insinuation that he was being edged out of the service because of his critical writings also seems far-fetched—it would have been more plausible a few years later.

22. *Memoirs* pp. 27–65 *passim*. Jay Luvaas *The Education of an Army* (U. of Chicago 1964: Cassell 1965) p. 378.

23. 8/301, 8/317. Liddell Hart to Captain J. M. Scammell 22 February 1923.

24. *Memoirs* pp. 37–38.

25. 'The Essential Principles of War and their application to the offensive infantry tactics of today' in *United Service Magazine* April 1920 pp. 30–44. See also Luvaas op. cit. pp. 378–380.

26. 'The "Man in the Dark" Theory of War' op. cit. *Memoirs* pp. 31–39.

27. Barrie Pitt *1918: the Last Act* (Cassell 1962) pp. 43–44. H. Essame *The Battle for Europe, 1918* (Scribner's: Batsford 1972) pp. 46, 102, 136ff., 181ff.

28. Maxse correspondence file 1920–21 *passim*. See also 7/1920/55 *et seq*.

29. 7/1920/110 Liddell Hart to Evetts 24 August 1920. There is no suggestion in the letters of Maxse, Dugan or Evetts that Liddell Hart's idea of the expanding torrent was familiar to them or 'old hat'.

30. 7/1925/7 'Early Record of Career' (holograph). See also *Memoirs* pp. 43–46; and Luvaas op. cit. p. 380.

31. 7/1920/56 Maxse to Liddell Hart 8 November 1919.

32. 'Suggestions on the Future Development of the Combat Unit: the Tank as a Weapon of Industry' in *R.U.S.I. Journal* November 1919 pp. 666–669. See also Luvaas pp. 380–381.

33. Note for example the few and brief references to artillery in *The Remaking of Modern Armies* (1927) especially pp. 22–23.

34. Luvaas p. 353. 'Plan 1919' is printed in Fuller's *Memoirs of an Unconventional Soldier* (Nicholson & Watson 1936) pp. 322–336.

35. Fuller to Liddell Hart 10 June 1920. Liddell Hart to Fuller 15 June 1920.

36. Liddell Hart to Fuller 16 and 31 January 1922. Fuller to Liddell Hart 19 January 1922.

37. *Memoirs* pp. 86, 89–91.

38. 'My Rating of Officers' 7/1925/2. Liddell Hart to Captain J. M. Scammell 22 February 1923. Liddell Hart to Fuller 11 March 1928. Earlier, Liddell Hart had told Fuller he considered the latter's lectures at the Staff College 'the most brilliant contribution to scientific military thought that has ever been made' (11 April 1923).

39. For example, Fuller to Liddell Hart 4 June 1923, 'He [Scammell] is right. I shall end by having my throat cut—it is inevitable.' Fuller's letters to Liddell Hart abound with defamatory remarks about his military superiors, which Mrs Fuller was liable to repeat at social gatherings. For example, on hearing that a despised general had been awarded the G.C.B., Fuller suggested that the initials must stand for 'Great Cretin Brotherhood'!

40. 7/1920/386 (typed copy, 23 pp.). Although readers of his *Memoirs* may consider that Liddell Hart had already enjoyed a certain amount of luck, he could write in 1920: 'I can safely say that everything I have done has been in spite of the luck being dead against me.' Ibid. p. 12.

41. 7/1921/68.

42. 'Some Personal Impressions' (4 typed pp.) 7/1920/32 (retyped in December 1929 but original clearly written in 1920).

43. 'The Development of the "New Model" Army. Suggestions on a Progressive, but Gradual Mechanicalization' in *Army Quarterly* October 1924 pp. 37–50. The keynote of this article was 'fire-power multiplied by speed': 'This speed, only to be obtained by the full development of scientific inventions, will transform the battlefields of the future from squalid trench labyrinths into arenas wherein manoeuvre, the essence of surprise, will reign again after hibernating for too long within the mausoleums of mud. Then only can the art of war, temporarily paralysed by the grip of trench warfare conditions, come into its own once more.'

44. *Memoirs* p. 75. Luvaas op. cit. pp. 383–384.

2

The Strategy of Indirect Approach, 1925–30

In the second half of the 1920s Liddell Hart achieved an international reputation as an outstanding commentator on military matters past and present. His own interests widened from a preoccupation with infantry tactics and the individual in combat to embrace history, strategy and a general theory of war. The main external impetus to a period of almost frenetic study and writing was supplied by his increasing disenchantment with the conduct of the First World War and a hardening conviction that the chief cause of the futile holocaust had been adherence to a false military doctrine, namely Clausewitz's interpretation of Napoleonic warfare.

In the decade following 1918, war-weariness together with some positive diplomatic achievements provided grounds of hope for the optimists who believed that the recent débâcle had been 'the war to end all wars'. The unprecedented international agreement on measures of naval disarmament and arms control at the first Washington Conference (1922) was followed by the Locarno Treaty (1925), which guaranteed the security of the frontiers of Western Europe, and by Germany's admission to the League of Nations (1926). The brief period of euphoria in the later 1920s was epitomized by the Kellogg–Briand Pact (1928), whose signatories undertook to outlaw war as an instrument of national policy. In Britain, defence preparations after 1919 were guided by the assumption that the country would not be involved in a major war for at least ten years.

Like most professional soldiers, Liddell Hart was not convinced that a new era of permanent peace was dawning. He was not a pacifist, and lacking faith in complete disarmament or the international renunciation of war as an instrument of policy, he strove rather to improve the future conduct of war on two levels. In practical, operational terms he became one of the foremost advocates of mechanization and mobility; while in theoretical terms he attempted to devise a counter to what he regarded as

Clausewitz's evil legacy in the form of the 'strategy of indirect approach'. These years (1925–30) produced a steady stream of publications in which he combined the 'lessons' of past military history (in which he firmly believed) with critical studies about the First World War and recommendations for the present and future. What needs to be emphasized at the outset is that in these years Liddell Hart was first and foremost a busy journalist who somehow made time for a remarkable amount of additional research and writing. Moreover, though consciously aware of the need for ruthless objectivity and 'scientific' detachment in the extraction of lessons from history, he was also very definitely a committed critic with a sense of mission. Consequently, his publications, though possessing some scholarly attributes, should be regarded primarily as polemics or tracts for the time. T. E. Lawrence, who shared Liddell Hart's antipathy to Clausewitz, offered an interesting observation on the former's endeavours in 1928:

> There is, in studying the practice of all decent generals, a striking likeness between the principles on which they acted: and often a comic divergence between the principles they framed with their mouths. A surfeit of the 'hit' school brings on an attack of the 'run' method: and then the pendulum swings back. You, at present, are trying (with very little help from those whose business it is to think upon their profession) to put the balance straight after the orgy of the late war. When you succeed (about 1945) your sheep will pass your bounds of discretion, and have to be chivvied back by some later strategist. Back and forward we go.[1]

In 1925 Liddell Hart published a little volume entitled *Paris, or the Future of War* (in a series called 'To-day and Tomorrow') which constitutes a landmark in the development of his general ideas on warfare. Though time has dealt harshly with some of his predictions, it remains one of his most interesting books.

His basic premise was that the Allies had followed an erroneous strategy in the First World War; namely the destruction of the enemy's armed forces in the main theatre of war. Ordinary citizens had paid a terrible price for this blunder, 'yoked like dumb, driven oxen to the chariot of Mars'. Worse still, he alleged, the orthodox school ('these military Bourbons') restored to authority had learnt nothing and forgotten nothing.

'New weapons would seem to be regarded merely as an additional tap through which the bath of blood can be filled all the sooner.' He feared that even in the new element air, forces would be harnessed to the Napoleonic theory, that is, directed towards the destruction of the enemy's air force. Only if this tradition could be broken would war (or international conflict generally) once again serve a meaningful purpose: 'A prosperous and secure peace is a better monument of victory than a pyramid of skulls.'

Clausewitz would not have dissented from Liddell Hart's view that the aim in war is 'to subdue the enemy's will to resist, with the least possible human and economic loss to itself', but the latter took a more original line in asserting that 'a highly organised state was only as strong as its weakest link'. If one section of the nation could be demoralized, he argued, the collapse of its will to resist would compel the surrender of the whole; this had been demonstrated by the last months of 1918. The function of grand strategy was to discover and exploit the Achilles' heel of the enemy nation (just as Paris, son of Priam, had killed the Greek champion—hence the punning title): in short, as a general rule one should strike against the enemy's most vulnerable spot rather than his strongest bulwark.[2]

How can the military weapon best be employed in war to subdue the enemy's will to resist? Liddell Hart's answer evinced a surprising faith in rationality in the face of recent evidence: armies and nations, he believed, were mainly composed of normal men, not of abnormal heroes, and once these realize the *permanent* superiority of the enemy they will surrender to *force majeure*. Armies, he then pointed out in terms entirely familiar to Clausewitz, suffered from a serious handicap in subduing the hostile will:

> Being tied to one plane of movement, compelled to move across the land, it has rarely been possible for them to reach the enemy capital or other vital centres without first disposing of the enemy's main army, which forms the shield of the opposing government and nation.

Like many contemporary thinkers in the 1920s, Liddell Hart found the solution in the aeroplane. Liddell Hart was trying out extreme air-power theories on Captain J. M. Scammell as early as the autumn of 1923. Unfortunately his paper has not survived, but its tenor can be appreciated from Scammell's sceptical comments:

> I can't quite agree that Air Power is going to absorb land and sea functions. Just as land power and sea power have their peculiar powers and also their inherent limitations, so air power has its necessary limitations. Radius of action may be enlarged within limits to be determined by the relation between lifting power, other necessary weights ... and weight of fuel. Design will affect fuel consumption, and the distribution of bases affect flexibility of operations; but the idea of planes in numberless quantities flying everywhere at will must be modified ... They [aircraft] cannot *occupy*. Their influence must be ephemeral. Also the weaker power is not helpless ... As for bombs and gas wiping out cities etc that seems to me to be rot.[3]

Liddell Hart at this time subscribed wholeheartedly to the view that air power would dominate future conflict. A modern state, in his view, is such a complex and interdependent fabric that it offers a target highly sensitive to a sudden and overwhelming blow from the air.

> Aircraft enables us to jump over the army which shields the enemy government, industry, and the people and so strike direct and immediately at the seat of the opposing will and policy. A nation's nerve system, no longer covered by the flesh of its troops, is now laid bare to attack, and, like the human nerves, the progress of civilization has rendered it far more sensitive than in earlier and more primitive times.[4]

He painted a terrifying picture based on a projection of the limited bombing experience of World War I into the future:

> Imagine for a moment that, of two centralized industrial nations at war, one possesses a superior air force, the other a superior army. Provided that the blow be sufficiently swift and powerful, there is no reason why within a few hours, or at most days from the commencement of hostilities, the nerve system of the country inferior in air power should not be paralysed ...
>
> Imagine for a moment London, Manchester, Birmingham, and half a dozen other great centres simultaneously attacked, the business localities and Fleet Street wrecked, Whitehall a heap of

> ruins, the slum districts maddened into the impulse to break loose and maraud, the railways cut, factories destroyed. Would not the general will to resist vanish, and what use would be the still determined fractions of the nation, without organization and central direction?

It would be folly, he concluded, to base the military forces on infantry when aircraft could reach and destroy cities such as Essen or Berlin *in a matter of hours*.[5]

How would the transcendent value of air power affect the two older services? Liddell Hart went so far as to assert that 'ultimately' the air would become 'the sole medium of future warfare'. He also made some sweeping assertions which show that he had as yet given little thought to the practical problems of air power, such as that it would be simpler to convert civil aircraft to military use than with any of the old-established arms, or that gas (described here as the 'ideal weapon') would be unaffected by the conditions or physical defects of the firer. The implication was that aeroplanes and gas would eventually provide a perfect combination in permitting the demoralization of the enemy with a minimum of physical destruction and death. He did, however, make the crucial perception, which ought to have modified some of his more striking statements, that:

> Where air equality existed between rival nations, and each was as industrially and politically vulnerable, it is possible that either would hesitate to employ the air attack for fear of instant retaliation.[6]

On the naval side, Liddell Hart confessed his lack of technical knowledge and contented himself with the remark that although the battleship would retain the sovereignty of the oceans for some time to come, submarines would tend to dominate in the narrow seas. He had little to say about the effects of air power on capital ships, but he did see very clearly that a hostile power (France was cast in that role in 1925) could deny Britain the use of the Mediterranean through the combined menace of submarines and shore-based aircraft. Britain was clearly dependent on the goodwill of the other and better-known 'Paris'.[7]

The last war, according to Liddell Hart, was the culmination of brute force; the next would witness 'the vindication of moral force, even in the realm of the armies'. Infantry must be provided with 'mechanical legs' to carry them to the battlefield, horse-drawn artillery must be replaced with motor-drawn or motor-borne guns, and the tank arms must be developed as the successor to the cavalry. With such a 'New Model' army able to operate independently of roads and railways, advances of a hundred miles in the day would be possible. He sketched exciting prospects if tanks were properly employed:

> Once appreciate that tanks are not an extra arm or a mere aid to infantry but the modern form of heavy cavalry and their true military use is obvious—to be concentrated and used in as large masses as possible for a decisive blow against the Achilles' heel of the enemy army, the communications and command centres which form its nerve system. Then not only may we see the rescue of mobility from the toils of trench-warfare, but with it the revival of generalship and the art of war, in contrast to its mere mechanics.[8]

This provocative essay received wide and generally favourable coverage in the international press. The most criticized aspect was Liddell Hart's championship of gas as a relatively benign yet effective weapon. Not surprisingly, however, his interpretation of the future of air power was also seriously challenged. For example, Generals Sir Ian Hamilton and Sir Frederick Maurice, writing in the *Westminster Gazette*, pointed out that gas and chemical warfare might not aid the superior civilization, and that at present air power could not paralyse a city in a few hours—though it might in the future.[9]

On the latter point, Liddell Hart had evidently not been convinced by the arguments of Captain J. M. Scammell, who had disputed his view that air power could ever be decisive by itself. The latter rightly pointed out that in the First World War 'aircraft raids did not break the will of the British or French. They strengthened it. Even worse harassing has often failed. Will-breaking is a psychological matter. The best means of really breaking the will to resist are:

1. A sharp, concentrated, dramatic stroke and
2. A kind, generous policy'.

Later, Scammell's former tutor at Oxford, Spenser Wilkinson, took up the same issue, pointing out that Liddell Hart's idea of by-passing the enemy's army to strike direct at his civil population was not new: it had been the normal practice before the rise of the modern state, whose prime duty was to protect its citizens, and was in fact utterly brutal. Moreover airmen were mistaken in thinking they would be able to bomb unimpeded and at will: antidotes would soon be found to the bomber aircraft and also to the tank. As to the future pattern of war in the air, Wilkinson wrote presciently: 'I don't think it possible that it will be (except for a possible brief period of transition) the rush of air forces to destroy the enemy resources. I think it will be a battle in the air in which the victor will have the beaten side at his mercy....'[10]

According to Liddell Hart, Sir Hugh Trenchard, then Chief of the Air Staff, was so impressed by *Paris* that he ordered a number of copies for distribution in the R.A.F. and also gave copies to his fellow Chiefs of Staff. Another Air Marshal wrote to the author that *Paris* 'is practically what is taught at the R.A.F. Staff College'.[11] It seems improbable that the R.A.F.'s strategic doctrine owed anything to the writings of the Italian aviator, Giulio Douhet, though he was the most forceful exponent of aerial bombing as the decisive instrument of war in the 1920s. Certainly Liddell Hart had not read him when he wrote *Paris*.[12]

This thesis, of aerial bombardment as a decisive method of war, though so prominent in *Paris*, was to receive little emphasis in Liddell Hart's later books. The reason may be in part that he came to appreciate the practical limitations of existing aircraft and their ability to locate and hit defined targets, but a more likely explanation is that the bombing theory conflicted with Liddell Hart's humanitarian instincts and overriding concern that future warfare should be limited. Strategic bombing was likely to be a clumsy, indiscriminate instrument inherently unsuited to the waging of limited war. In the Second World War Liddell Hart became a leading opponent of the Allied strategic bombing of Germany.

Liddell Hart's jaundiced view of Clausewitz and Napoleon (referred to in *Paris* as 'the Corsican vampire') will be discussed more fully later in this chapter, but it must be made plain here that he was grossly unfair to Clausewitz as regards making the enemy's army the main objective, and indeed he was unwittingly stating the Prussian writer's viewpoint in parts of *Paris*.

Spenser Wilkinson had vainly made this point in 1924 when he argued that Clausewitz 'quite understands that you must break the enemy's will, for which he suggests three processes: first to crush his army, and then to take his capital and, if that is not enough, to occupy his territory'.[13] Liddell Hart's retrospective anger over the strategy and tactics of attrition pursued on the Western Front is understandable, for there was in truth much to criticize and correct for the future, but he was somewhat less than scholarly or scientific in his search for the immediate causes of the deadlock and the complex process by which pre-war doctrine had been formed.[14]

Lastly, one may speculate as to how far the author had calculated the likely effects of his highly individualistic style: alliterative, abounding in colourful phrases, dogmatic and sharply iconoclastic. It is easy to see from the reviews that the style (and contents) delighted the general reader and those predisposed to criticize the high command, but more reflective soldiers and students of war were more likely to be irritated. Thus the anonymous reviewer in the *Army Quarterly* reacted crossly to the irritation and dogmatism displayed in 'this little tract' and reproved the author for failing to see that the methods of warfare adopted by Napoleon and others were often the best means available for destroying the enemy's will to resist.[15] Liddell Hart was already forcing history to yield the appropriate lessons.

In the later 1920s Liddell Hart interwove the results of his historical studies with lessons derived from the misconduct of the First World War to form his theory of the strategy of indirect approach. As a firm believer that lessons should be learnt from history, he was fond of quoting Bismarck's dictum, 'Fools say that they learn from their own experience. I have always contrived to get my experience at the expense of others.' This approach was manifest in his first historical work, a biography of Scipio Africanus entitled *A Greater than Napoleon* (1926). Here Liddell Hart brilliantly exploited the limited sources available in English to make Scipio an admirable vehicle for his own military views. As Jay Luvaas aptly observes: 'A vital and instructive biography ... this is perhaps one of his best books from a literary point of view although Liddell Hart himself later questioned whether it is not too "coherent and convincing".'[16]

The polemic intent was evident in what would now be called a 'gimmicky' title: Spenser Wilkinson and Fuller both warned him of the unwisdom of asserting that Scipio was 'greater than' Napoleon or any other general from a different era, but he remained addicted to 'league tables' and lists in order of merit. The former also criticized the forcing of historical lessons and gently exposed the author's ignorance of modern scholarship regarding Hannibal, while Fuller deplored the absence of references. Not surprisingly, the book was welcomed in Italy, but in praising it the *Giornale d'Italia* remarked that it was often polemical, 'written for the purpose of defending and demonstrating a thesis and has at times the force of panegyric'.

Captain J. M. Scammell, probably Liddell Hart's most forthright, constructive critic, warned the author of two stylistic faults: 'a tendency to moralise and point even obvious conclusions'; and a heavy reliance on metaphors etc. which brought him dangerously close to triteness. But as regards arrangement and treatment 'I consider that nothing short of genius is shown in the way in which you have made Scipio live and have laid bare as a unified conception his system of command in war. I know nothing like it anywhere ... There is something very refreshing about your book ... it holds the attention like a novel. You have outdone yourself'. Others singled out the didactic aspect for special praise: 'What seems so valuable,' wrote John Buchan, for example, 'is the way in which you perpetually illuminate the argument with modern instances.'[17]

A few examples must suffice to illustrate the didactic element in what was primarily a gripping story of brilliant generalship and statecraft. In discussing Scipio's limited ability to control the battle, the author remarked that aeroplanes were now an asset. Air power now also provided the means to emulate the Romans' deliberate slaughter of civilians. Unlike the leaders in 1914–18 at Loos and elsewhere, Roman generals understood that a breach had to be promptly widened. Scipio's strategy of invading Africa while Hannibal remained in Italy provided a cue for praise of the 'Eastern' strategy of indirect approach in World War I as against the Western Front obsession. As for peace terms: '202 B.C.–1919 A.D.! What moderation compared with Versailles. Here was true grand strategy—the object a better peace.' Finally Scipio had shown an understanding of war 'such as is just dawning on the most progressive politico-military thought of today'; he had translated understanding into action 'in a way that we may possibly achieve in the next great war—more probably, we shall be fortunate to get out of the physical rut by 2000 A.D.'.[18]

Great Captains Unveiled (1927) consisted of a collection of articles about Jenghiz Khan, Sabutai, Maurice de Saxe, Gustavus Adolphus, Wallenstein and Wolfe which had first been published in *Blackwood's* and subsequently in the United States. As in the study of Scipio, there is again a marked tendency to view the past in terms of the present; to seek 'a message and a moral'. In reviewing the book favourably Scammell remarked that 'these sketches are more than biographies—they are propaganda ... a good word is said on behalf of tanks on every appropriate occasion'. 'As a historian,' commented the *Observer* ironically, 'Captain Liddell Hart takes as his field the past, the present, and what is to be. With one foot on a tank and one on a testudo, why, man, he doth bestride the military world like a Colossus, and we petty men walk under his huge legs and wonder where on earth we are.'[19] Although some of his correspondents queried historical details and omissions (Fuller, for example, urged the importance of Maurice of Orange's tactical innovations), these were really beside the point: Liddell Hart was less concerned with making a contribution to historical understanding—he already held academic historians in low esteem[20]—than in influencing contemporary military theory and practice. This is not to say, however, that his own ideas were consciously imposed on history; rather that for him the study of history served to mould, confirm and sharpen views which he had originally derived from reflection on the present and the recent past.

In *Great Captains Unveiled*, then, the author used his subjects to speak out on questions of tactics, strategy and policy for the benefit of his contemporaries. Thus the Mongols with their emphasis on quality, their youthful generals, their battle-drill system and their use of the indirect approach exemplify the mobile style of warfare which Liddell Hart hoped to see revived in his own day. 'I enjoyed Jenghiz Khan,' wrote Sir Ivor Maxse, 'and especially the masterly infiltration into it of our platoon training effort!' The lesson was underlined that the Mongols' combination of mobility and offensive power could now be regained through the development of a doctrine based on tanks and aeroplanes. Saxe was portrayed as the antithesis of Clausewitz, favouring quality and mobility as against numbers. According to Liddell Hart, neglect of the former's ideas in World War I meant that 'the artist yielded place to the artisan'. Germany, in his view, was learning from her recent defeat: her new military doctrine embodied the supreme importance of mobility and exploitation of manoeuvre by even the smallest units of infantry. Gustavus Adolphus was portrayed chiefly as the reviver of cavalry: with the advent of a 'new mechanical charger', the tank, Liddell Hart foresaw the rebirth of the cavalry charge and the decisive warfare of the Great Captains. Wallenstein was praised for his grasp of the 'moral weapon': his ruthless campaigns had underlined the folly of the 'idealistic pacifism' which was rampant in 1627 and again prevalent in 1927. Wolfe was seen as a brilliant exponent of the 'British Way in Warfare' on the strategic plane and of the indirect approach on the tactical. His career showed that military genius and not mere competence decides the fate of nations. Yet in peacetime Wolfe had been an outspoken critic of the system and of his superiors; such a man could not rise to the top in the British Army in the 1920s.[21]

Liddell Hart's biography of Sherman occupies a unique place in his intellectual development. He was naturally drawn to the study of the American Civil War because of the similarity of the trench deadlock in Virginia in 1864–65 to that on the Western Front in 1914–18, although he saw both conflicts almost entirely in military terms. Furthermore Liddell Hart soon appreciated that in British military thought attention had focused too narrowly on the eastern theatre, largely through the influence of Colonel G. F. R. Henderson's famous biography *Stonewall Jackson*. Consequently, when invited to write a book on the war in 1928 he preferred the comparatively neglected Sherman to a financially tempting study of Lee, whose popularity in Britain as well as the United States would probably have guaranteed larger sales.

In Liddell Hart's brilliantly written study, based on a wide range of primary as well as secondary sources,[22] Sherman shone forth as the embodiment of the strategic indirect approach: the general whose operations in the West had really won the war in contrast to Grant's costly frontal assaults in Virginia, and whose methods pointed the way to the revival of mobile warfare. In the march to Atlanta, Sherman had been obliged to adopt a direct approach as a result of his dependence for supplies on a single railway line, but even so he had managed to manoeuvre his wary opponent (General Joseph E. Johnston) out of successive strong defensive positions by advancing 'in a wide, loose grouping or net'. In addition to the concept of the 'strategic net', the study of Sherman's campaign also taught Liddell Hart the value of 'alternative objectives' and the 'baited gambit'. Time and again the Confederates were lured into vain attacks on what were really well-constructed positions. In the famous march from Atlanta to the sea and then up through the Carolinas, Sherman, now freed from dependence on a single supply line, was able to practise a true indirect approach threatening alternative objectives so that the Confederates were never certain of his destination until it was too late to stop him. The imagery and concepts which found fuller expression in *The Decisive Wars of History* (1929) were derived more from Sherman than from any other of Liddell Hart's military heroes. The latter bore testimony to Sherman's pervasive and long-lasting influence in his *Memoirs*:

> While the concept of deep strategic penetration by a fast-moving armoured force had developed originally in my mind when studying the Mongol campaigns of the thirteenth century, it was through exploring the rival operations of Sherman and Forrest that I came to see more clearly its application against modern mass armies, dependent on rail communications for supply. Study of Sherman's campaigns also had a strong influence on other trends in my strategical and tactical thought. Any reader of my subsequent books, in the nineteen-thirties, can see how often I utilised illustrations from Sherman to drive home my points—particularly the value of unexpectedness as the best guarantee of security as well as of rapid progress; the value of flexibility in plan and dispositions, above all by operating on a line which offers and threatens alternative objectives (thus, in Sherman's phrase, putting the opponent on the 'horns of a dilemma'); the value of what I termed the 'baited gambit', to trap the opponent, by combining offensive strategy with defensive tactics, or elastic defence with well-timed riposte; the need

> to cut down the load of equipment and other impedimenta—as Sherman did—in order to develop mobility and flexibility.[23]

Liddell Hart's deductions from the American Civil War are of course intrinsically important for what he made of them, irrespective of the accuracy or depth of his historical scholarship. A full discussion of Liddell Hart's treatment of the Civil War would be inappropriate here, but since disagreement over the respective merits of Grant and Sherman led to a serious clash with Fuller, a few comments are necessary.[24] Sherman's advance to Atlanta was clearly a special case in that his weaker opponent had no intention of fighting a pitched battle (as his successor John B. Hood recklessly did) and therefore was prepared to retreat. By contrast, Lee could not afford to retreat from before the Confederate capital of Richmond or the communications centre of Petersburg, and his defensive tactics were also greatly assisted by the terrain. It was unfortunate that in praising his hero, Sherman, Liddell Hart found it necessary to denigrate his superior commander, Grant. Fuller made the sound point that Grant and Sherman were pursuing complementary, not rival strategies (a 'lesson' that also applied to the First World War). Fuller also felt that Liddell Hart had over-romanticized his ruthless hero and archly suggested 'A Greater than Jenghiz Khan' as a more fitting catch title. Liddell Hart continued to view the Civil War through the eyes of 'his' Sherman and was never prepared to change his excessively critical view of Grant.[25] On a wider view, it is characteristic of Liddell Hart's outlook that he should see the Civil War primarily as a competition between brilliant generals rather than a conflict between different types of economy and society.

In 1927 Liddell Hart had published another collection of his articles entitled *The Remaking of Modern Armies*, whose keynote was again 'mobility' and whose moral was that 'nothing less than rebirth can revive mobility in armies and in warfare. And without mobility an army is but a corpse—awaiting burial in a trench. Endowed anew with mobility through armour and mechanisation, armies may still play a great role.'

The most interesting section of the book is Part III, originally published as a series of articles in *The Daily Telegraph*, which contains a prophetic analysis of the post-war doctrines of the

French and German armies. Liddell Hart's acuteness as a military critic was never better displayed; already, in 1927, he saw 'the writing on the wall'. Germany, profiting from defeat, was placing great emphasis on manoeuvre, whereas France's doctrine was rigid in form and static in its vision of future war: tanks would be used mainly to protect infantry while artillery remained the queen of the battlefield. In a future war Liddell Hart feared that German quality would defeat French quantity:

> How will the French steam-roller fare if its enemy declines to roll ponderously against it, and instead, barring large areas with gas, concentrates its highly-trained forces for mobile thrusts aimed to dislocate the complex French war-machine? It is paradoxical that the French and Germans should have exchanged doctrines, in essence at least, the Germans embodying the Napoleonic tradition of mobility and surprise, while the French forsake their historic élan and swift manoeuvres in their devotion to fire-power.

France, as he correctly perceived, stood at the parting of the ways:

> If she continues on her present course, her Army will become a national militia, growing ever less effective owing to lack of expert instructors, as her old type of professional officers, devoted and highly educated, drop out from lack of adequate pay and prospects.
>
> What is the alternative course? My own conviction is that a smaller professional army, well paid, highly trained and mechanised, would by its quality give France more real security than her present quantity affords. I would go farther, and say that far from being a militarist menace, as is the delusion of ignorant foreign pacifists, France to-day has hardly adequate cover in her military insurance policy, and this security is shrinking to danger-point. The situation is serious because a secure and strong France is essential to the stability of European peace.[26]

By this time Liddell Hart had hardened in his view that the futile attrition of the First World War was fundamentally due to the prevalence of 'The Napoleonic Fallacy', namely that the prime objective in war should be the enemy's main army. He reprinted a chapter with that title in *The Remaking of Modern Armies* which had already appeared in *Paris* and would be published yet again in *The Ghost of Napoleon*. In a review article entitled 'Killing No Murder', and in private correspondence, Spenser Wilkinson vainly tried to convince Liddell Hart that he was misrepresenting Clausewitz, who had never deluded himself that the enemy's armed forces were the real objective.[27] Indeed Wilkinson demonstrated, convincingly in my view, that Liddell Hart was actually repeating Clausewitz in most of his main arguments. Liddell Hart, wrote Wilkinson, was in fact a member of the orthodox school and was propounding a paradox: he was after the old will o' the wisp, victory without battles or bloodshed. Though himself rather dogmatic in arguing that *even more* concentration on the Western Front would have been a more effective strategy, Wilkinson achieved a rare insight in perceiving that the basic reason for the War's 'total' nature was the involvement of a number of great powers all believing their vital security to be at stake and all prepared to pay a high price for it. In this *he* showed himself to be a good disciple of Clausewitz.

Lastly, Wilkinson made three excellent points against Liddell Hart's view of aerial bombing of cities as the way to avoid the enemy's army: first, defences were in their infancy and would doubltess improve in time; second, 'the lessons of history' suggested that such attempts at terrorism could actually strengthen resistance and would be ineffective unless the enemy's armed forces were also defeated; and, third, in the bombing theory it was crucial to strike the first blow—a hopeless prospect 'as it is quite certain that no British Government will take the initiative in making war'. The truth of this last point, despite all the arguments of the offensive bombing school, was completely borne out by the British Government's attitude to strategic bombing in the opening months of the Second World War.

One must conclude that Liddell Hart was so emotionally involved in attacking the inept conduct of the First World War and its legacy that he was unable to approach its more general causes with detachment. Instead he found a plausible scapegoat in what he mistakenly believed to be Clausewitz's notion of strategy. This confusion is not at all surprising; nor does it necessarily invalidate the operational lessons he derived from the First World War. In terms of air warfare he lacked technical knowledge and personal experience, and shared many of the false assumptions of the 'strategic bombing' advocates. But as we have seen, his understanding of the likely outcome of a future war between France and Germany, judged by their respective military doctrines in the later 1920s, was brilliantly accurate.

Captain Scammell, who also dissented from Liddell Hart in some of his extreme positions, including his view of future air warfare, probably hit the nail on the head in accounting for his friend's provocative style and opinions. Scammell believed that if confronted by a concrete problem in war Liddell Hart would be the first to qualify his conclusions:

> His object [he wrote in reviewing *The Remaking of Modern Armies*] is less to describe actual or impending conditions than it is to awaken the military from their lethargy and the public from its apathy. As the comfortable but deceptive drowsiness creeps over a victim of extreme cold, he must be cuffed and kicked into activity ...[28]

As for Clausewitz, Spenser Wilkinson was almost alone in Britain at that time in fully understanding that his prime concern was neither to formulate a system nor to provide rules and precepts for the guidance of generals, but rather to think out the nature of war both in its general aspect and in its various phases.[29]

The chief interest of *Reputations* (1928), a preliminary assessment of ten not-so-great captains of the First World War, is that it shows Liddell Hart's attitude in a midway position between the uncritical adulation of 1916 and the increasingly severe judgement displayed in *The Real War* (1930) and subsequent revised and expanded versions.

First, however, a word about the depth and thoroughness of Liddell Hart's researches into the military history of the First World War. Considering that he was first and foremost a busy journalist, his research was extremely thorough as far as it went, but he hardly ever visited archives and only occasionally saw the original documents which for academic historians comprise the most important source. His lack of time to study in the archives was not such a fatal drawback then as it would be now, since official documents were not accessible to researchers until fifty years after the event. In any case he was able to circumvent this barrier to some extent by indefatigable correspondence and discussions with the official historians. As regards foreign languages he was able to work easily only in French and had to rely on translations of extracts from important publications in German and other languages. This imposed a particularly severe handicap on his understanding of the war on the Eastern Front, which he was forced to view mainly through translations of German generals' accounts. Thus Jay Luvaas exaggerates somewhat in claiming that behind each of his volumes on the First World War 'lies conscientious research among the documents [*sic*], diaries, and memoirs, lengthy conversations with the actors, and voluminous correspondence with those in a position to shed fuller light on the men and events under examination'.[30] A good example of his ability to build up a convincing account with scarcely any documentary evidence may be found in *Reputations*, where information for the chapters on the American generals Liggett and Pershing was gleaned largely from correspondence with Captain Scammell. Wherever possible, however, he took great care in cross-checking such second-hand evidence.

In method he resembled the maker of a mosaic, using every scrap of his own work, including obituary notices, gradually to build up and refine his ideas so that eventually a clear picture emerged. This habit of endless overlaying and repetition (or 'interweaving' as he euphemistically called it) was probably in part necessary to earn a living, but it often irritated his reviewers and poses problems for anyone who now wishes to establish the chronological order in which his ideas occurred. It must also be mentioned on the debit side that once he had made his mind up and recorded his view in a striking phrase or paragraph, he was apt to repeat it without reappraising its validity in the light of new evidence. This habit sometimes gave unnecessary hostages to his critics.

By the late 1920s published records (by the generals and their admirers) and private revelations (from such informed sources as Brigadier-General Sir James Edmonds, the chief British official historian of the war) added to Liddell Hart's apprehension that current Army leadership was still too complacent and out of date in its thinking. This caused his scale of judgement to tilt increasingly against those responsible for the conduct of the First World War. Amusing evidence for this is a list of possible titles he drew up in 1930 for his history of the 1914–18 war which included: 'No Napoleon', 'The Conflict of Nations', 'The World

Folly', 'The World in Wonderland' and 'The Headless Monster'. Similar disenchantment is evident in some 'reflections' at the end of his 1930 diary, in particular the note: 'Strange, yet very common, especially among soldiers is the mentality which regards the exposure of rottenness as worse than the rottenness.'[31] In 1922 he had described Field Marshal Sir Henry Wilson, extravagantly, as along with Colonel Henderson 'the only real military genius thrown up by us for the past century'. After the publication of Wilson's *Life and Diaries* in two volumes in 1927 he could write 'Never has any man so condemned himself'. The diaries exploded his reputation and 'stamped him as of third-rate judgement'.[32]

In *Reputations* his treatment of Haig was balanced and even generous, concluding that 'as a great gentleman, also in the widest sense, and as a pattern of noble character, Haig will stand out in the roll of history, *chevalier sans peur et sans reproche*, more spotless by far than most of Britain's national heroes'. By 1936 the heroic note had disappeared; now 'Haig was an honourable man according to his lights ... But he was not honest enough by true standards.' This and later more scathing criticisms of Haig's generalship and character, particularly in the Passchendaele campaign, are in my view justified by the evidence, but the critical tone was doubtless also due in part to his sense of waging an unending struggle against complacency, conservatism and the falsification of history: 'the men have gone', as he succinctly put it, 'the system remains'.[33]

The military doctrine or philosophy most closely associated with Liddell Hart's name, the Strategy of Indirect Approach, first found full expression in 1929 in a volume entitled *The Decisive Wars of History*. The main conclusions, which he summarized at length in his *Memoirs*, were that:

> More and more clearly has the fact emerged that a direct approach to one's mental object, or physical objective, along the 'line of natural expectation' for the opponent, has ever tended to, and usually produced, negative results. The reason has been expressed scientifically by saying that while the strength of an enemy country lies outwardly in its numbers and resources, these are fundamentally dependent upon stability or 'equilibrium' of control, morale and supply. The former are but the flesh covering the framework of bones and ligaments.

To move along the line of natural expectation is to consolidate the opponent's equilibrium, and by stiffening it to augment his resisting power. In war as in wrestling the attempt to throw the opponent without loosening his foothold and balance can only result in self-exhaustion, increasing in disproportionate ratio to the effective strain put upon him. Victory by such a method can only be possible through an immense margin of superior strength in some form, and even so tends to lose decisiveness. In contrast, an examination of military history, not of one period but of its whole course, points to the fact that in all the decisive campaigns the dislocation of the enemy's psychological and physical balance has been the vital prelude to a successful attempt at his overthrow. This dislocation has been produced by a strategic indirect approach, intentional or fortuitous ...

The art of the indirect approach can only be mastered, and its full scope appreciated, by study of and reflection upon the whole history of war. But we can at least crystallise the lessons into two simple maxims, one negative, the other positive. The first is that in the face of the overwhelming evidence of history no general is justified in launching his troops to a direct attack upon an enemy firmly in position. The second, that instead of seeking to upset the enemy's equilibrium by one's attack, it must be upset before a real attack is, or can be successfully, launched ...

Mechanised forces, by their combination of speed and flexibility, offered the means of pursuing this dual action far more effectively than any army could do in the past.[34]

As was his lifelong habit, Liddell Hart circulated the manuscript widely before publication for criticism and comment. T. E. Lawrence made a perceptive observation:

You establish your thesis: but I fear that you could equally have established the contrary thesis, had the last war been a manoeuvre war, and not a battle war. These pendulums swing back and forward. If they rested still that would be absolute truth: but actually when a pendulum stands still it's that the clock has stopped, not that it has achieved absolute time![35]

Major-General (later General Sir) W. H. Bartholomew, one of the best minds on the General Staff in the inter-war period, commented:

You are out to prove a theory by exemplification—always danger-

> ous, as one is so apt to forget to put the other side. Are you satisfied that you cannot be assaulted on that ground? Much as I like your book, I would have liked it better still if you had not used the words 'indirect approach' so much, but had allowed the value of the indirect approach to emerge. You are really out to prove that surprise, which throws the enemy off his balance as well as secures a position which makes the enemy's situation dangerous, is the best weapon in war. There is no virtue in an indirect approach unless it secures this end, and I feel that you have rather glorified the method at the expense of . . . the principle.[36]

Fuller likewise challenged the notion of the indirect approach as a cure-all: 'The object is to defeat the enemy and if this can be done by a direct approach so much the better.' In Captain Scammell's opinion the book proved that Liddell Hart was a genius. He pointed out that other students of war, including Camon, Colin and Boucher, had tried a similar approach and reached more or less the same conclusions. But Scammell's mentor, Spenser Wilkinson, remained sceptical. In 1932 he wrote to Scammell: 'I think Liddell Hart is a little too much the slave of his own theories which he makes into dogmas. But he is clever and knows a good deal.' And in another letter he put the Clausewitzian view: 'War is always an affair of State and part of the changes in war are changes in the constitution of the State. There never were and never will be recipes for victory.' Sir John Fortescue was numbered among Liddell Hart's outright critics: 'For all his study,' he wrote in reviewing the book, 'Liddell Hart's work is superficial and he is vague on the indirect approach and what is "decisive".'[37]

In terms of methodology the historical foundations of the 'indirect approach' were indeed far from secure. Liddell Hart's approach to history was intuitive and eclectic rather than, as he liked to believe, 'scientific'. In studying the book one feels that his method is largely self-fulfilling. Few great battles or wars have been won without some kind of subtlety, surprise or innovation and since Liddell Hart allows that the 'indirectness' can be strategic, tactical, psychological and sometimes even 'unconcious', he comes extremely close to a circular argument: by his definition a 'decisive victory' is an event which is secured by an 'indirect approach'. The method is unscientific in other respects: whole centuries, areas of the world and significant wars are omitted; and one would also expect more cases than he cites of 'indirect approaches' failing (Napoleon's abortive invasion of Egypt, for example) or direct approaches succeeding (tactically, for example, at Hastings and Blenheim). Perhaps most serious of all, Liddell Hart was pre-Clausewitzian in treating warfare and generalship almost entirely in isolation from their political, social and economic contexts. In other words, the reasons why warfare produced 'decisive results' in one era and not in another might have comparatively little to do with generalship. Liddell Hart was too emotionally involved in his particular view of the First World War, for example, to see analogies with the American Civil War, which became a war of attrition despite the presence of brilliant generals.[38]

Spenser Wilkinson, perhaps Liddell Hart's best-informed critic, suggested that his study was doctrinaire rather than historical: 'By that I mean that you set out to teach the dead generals how much better they would have done had they been imbued with your views of the indirect approach, whereas I think the historical method consists in finding out in each case, as far as possible, how it was that they acted precisely as they did ... My idea is that most of those great commanders very well knew their business and did the best it was possible in the conditions....' He indicated a number of important cases (such as Moltke's campaigns in 1866 and 1870) where Liddell Hart's interpretation was, to say the least, dubious. It is interesting that in his reply Liddell Hart denied any claim to originality in his general concept of the indirect approach:

> How could it be, as it is simply a way of crystallising the most successful methods of the great captains. As a thesis, the only newness one would claim is in degree and in emphasis. In degree largely by dwelling on its psychological effects rather than on its merely geometrical aspect. In emphasis because even Colin ... rather obscures its importance by making it appear incidental rather than fundamental, besides treating it mainly from the standpoint of tactical manoeuvre.[39]

At this distance in time it is easy to see that Liddell Hart was motivated only to a small extent by disinterested historical curiosity and to an overwhelming degree by his passionate reaction against the futile slaughter of the First World War and his concern to avoid a repetition of similar methods in the future. Thus what mattered far more than his historical accuracy and fairness—as indeed most of his contemporary readers seemed to realize—was the validity of his general thesis to present conditions and its practical relevance to future warfare. He expressed his overriding concern very clearly in a letter to Hugh Cudlipp in November 1940:

> What I constantly assailed for years was the orthodox military faith in the direct offensive i.e. the frontal attack against an enemy firmly posted in position—above all, if any such attack were made with the methods and means of the last war, with its ponderous artillery preparation. The method which I ceaselessly advocated was that of the 'indirect approach', in any form that would achieve surprise, while the means which I insisted on as necessary was the combination of the aeroplane and the tank.
>
> All this was anathema to the orthodox, who ruled the French and British armies. In 1932 I succeeded in getting a War Office Committee—which was appointed to investigate the lessons of the last war, and to see whether they were being adequately applied—to adopt my axiom that 'no attack in modern war is feasible or likely to succeed against an enemy in position, unless his resisting power has already been paralysed either by (a) some form of surprise, or (b) preponderating fire, powerful enough to produce the effect of surprise'. I also succeeded in getting the Committee to lay down a number of my other points—about methods of surprise, the necessity of armoured forces for a break-through, the value of the counter-offensive, and the vital importance of the indirect approach, etc. The report of this Committee, was, in effect, a devastating exposure of our military incompetence in the last war. But, although it had been originally intended that it should be published for the benefit of the rising generation, it was kept confidential through the influence of Sir Archibald Montgomery-Massingberd, who had become Chief of the Imperial General Staff by the time it was ready for publication. The result was that the British army continued to train for an out-of-date type of offensive that had no possible chance of success, and the French Army was still worse.[40]

This and many similar statements in his Papers show conclusively that Wavell was wide of the mark in his humorous suggestion: 'With your knowledge and brains and command of the pen, you could have written just as convincing a book called the Strategy of the Direct Approach.'[41] We have seen Liddell Hart putting the best case for the infantry arm in the early 1920s when already a convert to the greater potential value of tanks, and on most issues he retained a remarkable capacity to argue the pros and cons on both sides. But as regards the theory of indirect approach he had become what would now be called a 'committed' writer, still open-minded on matters of detail but adamant on this fundamental assumption in his philosophy.

Whatever its shortcomings from the viewpoint of scholarship, the Strategy of Indirect Approach can be strongly defended as an educational doctrine. There was a great deal to be said for encouraging a new generation of officers to think for themselves, and in particular to think in terms of achieving success by surprise and superior mobility; to value intellect and professional skill more than tradition and seniority; and to make the fullest use of science and technology to minimize casualties. In these aims Liddell Hart was only stating more eloquently and forcefully the aspirations of a generation of the more progressive-minded officers who had survived the holocaust of the First World War; hence his voluminous correspondence with such men as Fuller, Gort, Wavell, Montgomery, Ironside, Burnett-Stuart, Tim Pile, Hobart, B. D. Fisher, Henry Karslake and many more. To sum up, as a general stimulant to thought (or, as Wavell put it, a 'mental irritant') Liddell Hart's books, articles, letters and discussions gave marvellous value. Thoughtful officers were the chief but not the only beneficiaries. 'I always find that I learn from your conversation or writings,' the classicist and philosopher Professor Gilbert Murray wrote in 1941, 'and even when I do not entirely agree, you always make me think. You certainly do a valuable service in standing up to the flood of wishful thinking that war generates.'[42]

By 1930, when still aged only thirty-five, Liddell Hart had achieved an international reputation as a military commentator, historian and publicist. His appointments as military correspondent of *The Daily Telegraph* and military editor of *Encyclopædia Britannica* provided firm bases from which he sallied forth into history and current military issues. It was in the latter field, as a brilliant journalist leading the struggle for mechanization, that he probably achieved his greatest influence in the

1920s and 1930s. This struggle he has chronicled in great detail in the first volume of his *Memoirs*.

Friends and supporters were delighted at his rapid success, while others, who felt themselves to be the targets of his criticism, grumbled at this upstart Captain who presumed to tell the Generals their business. As Fuller reminded him, 'People are jealous of you. Look at your position [in] 1922 and 1927. In 1922 you had to say "Yes Sir" and "No Sir" to a twopenny halfpenny Captain; now you can put the wind up the Army Council.'[43] Though sounding excessively egotistical (but in accord with his belief that everyone ought to give themselves full value as he himself did), a note Liddell Hart made for John Buchan in 1932, when the latter had hopes of getting him appointed Deputy Secretary of the Committee of Imperial Defence, is true enough:

> During recent years many of the leading statesmen, soldiers, historians and newspapers of the world have delivered themselves of remarks to the effect that one is 'the greatest living thinker and writer on war'. Frequently [he added with unconscious irony], the remark has been 'the greatest since Clausewitz'.[44]

As early as 1926, his former Cambridge tutor, Geoffrey Butler, now a Member of Parliament and Parliamentary Private Secretary to the Air Minister, referred to him publicly at a College reunion as 'very distinguished' and 'a great man', and spoke of his giving the prestigious Lees Knowles Lectures at Trinity College, which in fact he did in 1933. In 1931 he was fulsomely praised in the United States Congressional Record where, according to his own notes, he was 'recognised as one of the most productive military minds of our generation'.[45]

This reference to his productivity was entirely deserved. His output of serious work amazed admirers such as John Buchan and J. M. Scammell, and eclipsed that of possible rivals (except Fuller) by its remarkable range and general high quality.

True, in terms of academic scholarship his historical works possessed obvious shortcomings, but with the exception of the elderly Spenser Wilkinson, who had retired in 1925, there were few if any university teachers in Britain at that time qualified to sit in judgement on him. Indeed, while rarely the product of original research in the strictest sense, Liddell Hart's books possessed obvious merits. They were based on wide reading supplemented wherever possible by interviews and correspondence, and their accuracy in detail was seldom successfully challenged. Not least important, Liddell Hart was a gifted writer; if he sometimes appeared to be carried away by his own brilliant phrases and images he nevertheless possessed the vital quality—particularly for an un-bookish military clientele—of readability. Again, with the honourable exception of Fuller, by 1930 there was no other British military writer who could emulate Liddell Hart in versatility, breadth of contemporary interest, independence of mind and courage in expressing unorthodox views.

Beyond all these qualities, Liddell Hart was already revealing, especially in his correspondence, that personal ambition for power and influence—though real enough—was transcended by a sense of mission. As this chapter has shown, he regarded himself as a crusader against the dangerous legacy bequeathed to the British military system by the strategy and tactics of the First World War. In the Strategy of Indirect Approach he had attempted to formulate an opposing general theory of war to that exemplified in the attrition campaigns of the Western Front. In the early 1930s he was to produce a more specific remedy for national security in his doctrine of 'The British Way in Warfare'.

Notes

1. *T. E. Lawrence to His Biographers Robert Graves and Liddell Hart* (1963) Lawrence to Liddell Hart p. 4.
2. *Paris, or the Future of War* (1925) pp. 12–13, 20–27.
3. J. M. Scammell to Liddell Hart 23 October 1923; see also *idem* 19 November 1923. Liddell Hart's belief in the pre-eminence of air power and his denunciation of Clausewitz for the evil effects of his doctrine in 1914–18 are combined in an article entitled 'The Napoleonic Fallacy: the Moral Objective in War' written in June 1924 and published in *Empire Review* May 1925.
4. *Paris* op. cit. pp. 42–43.
5. Ibid. pp. 46–48.
6. Ibid. pp. 56–61.

7. Ibid. pp. 64–68.

8. Ibid. pp. 79–85.

9. *Westminster Gazette* 3 July 1925. Newspaper and magazine reviews of Liddell Hart's early books are preserved in Extracts Album No. 8 (Liddell Hart Papers).

10. J. M. Scammell to Liddell Hart 19 November 1923. Spenser Wilkinson to Liddell Hart 14 and 28 August 1928.

11. *Memoirs* p. 99; Luvaas *The Education of an Army* p. 401n (U. of Chicago 1964: Cassell 1965). The earliest reference I can find to Trenchard's distribution of copies of *Paris* is Liddell Hart to Hankey 29 December 1933 but Liddell Hart's informant may well have been Squadron Leader C. G. Burge, Trenchard's Personal Assistant.

12. On Douhet's lack of influence in Britain see Robin Higham *The Military Intellectuals in Britain, 1918–1939* (Rutgers U.P., N.J. 1966) pp. 257–259. Liddell Hart's copy of Douhet's *Command of the Air* (1943) is unmarked.

13. Spenser Wilkinson to Liddell Hart 14 June 1924.

14. Cf Peter Paret 'Clausewitz and the Nineteenth Century' in M. Howard (ed.) *The Theory and Practice of War* (Cassell 1965: Praeger 1966) pp. 23–41. In the second half of the nineteenth century it was 'a Jominian rather than a Clausewitzian attitude that dominated military thinking' (*ibid.* p. 31).

15. *Army Quarterly* October 1925. The reviewer was almost certainly Spenser Wilkinson.

16. Luvaas op. cit. pp. 388, 394–395.

17. Fuller to Liddell Hart 16 August 1926. Spenser Wilkinson's review and comment on Liddell Hart's letter in the *Manchester Guardian* 9 and 17 December 1926. In the latter Wilkinson wrote 'I still think it futile to attempt to arrange the great captains in order of merit, and if I had favourites they would hardly be those chosen for me by Captain Liddell Hart'. *Giornale d'Italia* 29 December 1928. John Buchan to Liddell Hart 26 October 1926. Scammell to Liddell Hart 30 January 1927.

18. *A Greater than Napoleon: Scipio Africanus* (1926) pp. 34–37, 42–43, 91, 97, 152–153, 161–162, 189n, 194, 264.

19. Scammell in the *San Francisco Chronicle* 15 January 1928. *Observer* 13 November 1927. The latter's reference to Colossus and his 'huge legs' was amusingly apt in view of Liddell Hart's height—6ft 4in.

20. In 1926 he recorded his low opinion of 'cloistered dons' after meeting the 'fossils' at the Royal Historical Society; and in 1931 he reflected after meeting Maurice Bowra (who had been at prep school with him) and other dons, 'All graduates ought to go out into the world for ten years before becoming dons. Strange mixture of first class brains without experience' (11/1926/1 11 February; 11/1931/1 2 March). It is doubtful if the latter gibe was justified, for Bowra and his contemporaries, unless medically unfit, had been 'out into the world' between 1914 and 1918. Indeed Bowra, whose father was an official in the Chinese customs service, had travelled extensively even before 1914.

21. *Great Captains Unveiled* (1927) pp. 8–12, 20, 30–34, 40–44, 63, 118–120, 176–177, 209, 246–247, 273–274. Sir Ivor Maxse to Liddell Hart 3 October 1927.

22. *Sherman: Soldier, Realist, American* (New York 1929). In his *Memoirs* (pp. 168–169) Liddell Hart describes the preparation of his study of Sherman which, apart from *The Tanks*, was probably his most thoroughly researched book. In comparing the way in which previous authors misused or ignored the documents printed in the *Official Records of the Union and Confederate Armies* he discovered that some authors had copied their reference notes from other books including errors in quoting sources. He concluded that 'too much so-called research is slipshod, and the habit of giving footnote references is often a snare and delusion merely intended to impress the reader'. He used this dubious argument to defend the sparsity of his own footnotes for the remainder of his life.

23. *Memoirs* p. 166. For an excellent summary of the significance of Sherman in the development of Liddell Hart's ideas see Luvaas op. cit. pp. 398–400.

24. After a lengthy correspondence in 1929 in which Fuller had made some excellent criticisms of Liddell Hart's *Sherman*, the former lost his patience on being told he was a poor historian and addressed his reply on 12 July to 'Dear Lord God Almighty!'. On 25 November he wrote 'Your letter . . . is one of the most extraordinary examples of self-adoration I have cast an eye on for a long time . . . Because my Grant does not coincide with your Sherman, it is not Sherman who is injured but your colossal vanity. Like some literary Pope twice do you warn me of dire results to come—as if you were a judge and I a felon.' Liddell Hart had the good sense to reply disarmingly. On the respective merits of Grant and Sherman see also Wilkinson to Liddell Hart 28 December 1928.

25. Personal recollections of conversations with Liddell Hart.

26. *The Remaking of Modern Armies* (1927) pp. 250, 276.

27. Spenser Wilkinson 'Killing No Murder: an Examination of some new Theories of War' in *Army Quarterly* October 1927. On 18 January 1928 Wilkinson wrote to Scammell 'Liddell Hart is a keen fellow whom I am disposed to like. But he writes too much and is in a hurry. The right way to get there (wherever he is going) is to go quietly. Slow and sure' (Scammell file in Liddell Hart Papers).

28. Scammell in *San Francisco Chronicle* 14 October 1928.

29. For the tendency to misunderstand Clausewitz's attitude to war see P. Paret's essay in *The Theory and Practice of War*, op. cit. Liddell Hart sometimes states that he is only criticizing those who misread Clausewitz, but clearly he too concentrated almost exclusively on the operational sections of *On War*.

30. Luvaas op. cit. pp. 390–391.

31. 11/1930/23, 11/1930/1.

32. Liddell Hart to Scammell 24 June 1922. Luvaas p. 391. Sir C. E. Callwell *Field Marshal Sir Henry Wilson: his Life and Diaries* (2 vols, Cassell: Scribner's 1927). Another of the numerous books which revealed the less heroic side of World War I was Brigadier-General John Charteris *At G.H.Q.* (Cassell 1931).

33. *Reputations* (1928) p. 123. *Thoughts on War* (1944) pp. 149–150. *The Real War* (1930) pp. 361–367. For his later and more critical views see Liddell Hart

'The Basic Truths of Passchendaele' in *R.U.S.I. Journal* November 1959 pp. 1–7. I am grateful to Adrian Liddell Hart for pointing out the moral emphasis in his father's criticism of Haig, Henry Wilson and others. Curiously, he does not seem to have applied the same criterion to Lloyd George, whom he closely assisted in the preparation of his *War Memoirs* in the early 1930s.

34. *Memoirs* pp. 162–165. Luvaas pp. 397–398.

35. *T. E. Lawrence to His Biographers* op. cit. *Liddell Hart* p. 6.

36. Major-General W. H. Bartholomew to Liddell Hart (n.d.; 1928 or 1929).

37. Fuller to Liddell Hart 19 June 1929. Scammell to Liddell Hart 9 January and 11 February 1930. Spenser Wilkinson to Scammell 9 February and 30 June 1932. (See Scammell file, Liddell Hart Papers.) Sir John Fortescue in the *Observer* 22 September 1929.

38. It could be said that Liddell Hart revised without fundamentally changing Henderson's view of the Civil War by focusing attention on Sherman rather than Lee and Jackson. Had McClellan's campaigns in Virginia in 1861–62 succeeded they could have been cited as a brilliant example of indirect approach. Some American historians have suggested that the Union's early operations failed precisely because they were so indirect. See, for example, T. Harry Williams' essay 'The Military Leadership of North and South' in David Donald (ed.) *Why the North Won the Civil War* (Louisiana State U.P. 1960: Collier Books 1962).

39. Wilkinson to Liddell Hart 28 December 1928. Liddell Hart to Wilkinson 31 December 1928.

40. Liddell Hart to Hugh Cudlipp 15 November 1940.

41. Wavell to Liddell Hart 5 January 1935 (cited by Luvaas op. cit. p. 421). An American admirer of Liddell Hart, Admiral J. C. Wylie nevertheless noted that the strategy of indirect approach had been insufficiently defined: 'Indirection, in and of itself, is not necessarily something to be sought after and will not necessarily produce any result but diffusion!' J. C. Wylie *Military Strategy: a General Theory of Power Control* (Rutgers U.P., N.J. 1967).

42. Gilbert Murray to Liddell Hart 6 October 1940.

43. Fuller to Liddell Hart 5 December 1927.

44. Liddell Hart's typed copy of note requested by John Buchan October 1932 (John Buchan file).

45. Memoranda 11/1926/1 6 and 7 March. See also 29 March for Lord Esher's praise of Liddell Hart. 11/1931/1 Diary 13 February. For the Congressional Record Vol. 74 no. 22 10 January 1931 see 13/5 file 1. In 1931 Liddell Hart had good reason to consider he had been unjustly deprived by a prejudiced Council of the R.U.S.I.'s Chesney Gold Medal, awarded for outstanding contributions to military history (see his Diary for 1931 *passim*). He and Fuller belatedly received their Gold Medals together in 1963.

3

The British Way in Warfare, 1930–34

Although, in the first half of the 1930s, Liddell Hart persisted in his indefatigable efforts on behalf of the reform of the British Army, and also played an active part in discussion on disarmament, this chapter will concentrate on his broader strategic theories: in particular, his formulation of 'the British Way in Warfare' and its corollary, the denunciation of Continental mass warfare in *The Ghost of Napoleon*. The task of the next chapter will be to examine the practical application of Liddell Hart's ideas to Britain's defence policy in the later 1930s and to assess his influence. The present chapter will conclude by revealing something of his private preoccupation with 'the truth' and the deep pessimism about getting soldiers to accept it that features so prominently in his files for this period.

Liddell Hart first made public his notion of a specifically British 'Way in Warfare' in January 1931 in a lecture at the Royal United Service Institution entitled 'Economic Pressure or Continental Victories'. In that year the Manchurian crisis provided the first clear indication of the impotence of the League of Nations; in the following year hopes of international agreement on disarmament faded at Geneva; and early in 1933 the spectre of European war reappeared when Hitler became Chancellor of Germany.

In Britain the armed services had suffered severely from Treasury stringency throughout the lean years of the 1920s, reaching their nadir in the economic crisis of 1931–32. Military planning and expenditure had been governed since 1919 by the Cabinet's Ten Year Rule, which laid down that no major war could be expected during the next ten years and that no expeditionary force was necessary. In 1928 Churchill, as Chancellor of the Exchequer, had put the Ten Year Rule on a movable basis so that the date when readiness for war might be necessary never came any closer. Such assumptions may have been justified immediately after the end of World War I, but they were blatantly invalid by 1931. In March 1932, on the urgent recommendation of the Chiefs of Staff, the Cabinet cancelled the Ten Year Rule but in doing so warned 'that this must not be taken to justify an expanding expenditure by the Defence Services without regard to the very serious financial and economic situation which still obtains'. An increase of only £5 million in defence expenditure was granted for 1933: a drop in the bucket in view of the extent to which all three services had been allowed to run down.[1]

Towards the end of 1933 the Chiefs of Staff's annual Review warned that Germany was rearming and within a few years would again become a formidable military power. At any time within the next three to five years Britain might be obliged to intervene militarily on the Continent to fulfil her obligations under the Covenant of the League of Nations or the Locarno Treaties. At present, the Chiefs of Staff pointed out, the Army could provide at most two divisions which could not be reinforced for many months.[2] Also, they argued, it would be unrealistic to count on tranquillity elsewhere. In fact, the Review gloomily concluded, 'should war break out in Europe, far from our having the means to intervene, we should be able to do little more than hold the frontiers and outposts of the Empire during the first few months of the war'. This Review led directly to the setting up of the Defence Requirements Committee under the Committee of Imperial Defence to prepare a programme for meeting the worst deficiencies in the armed services over the next five years. Its chairman was Sir Maurice Hankey, Secretary of the Committee of Imperial Defence, and its members comprised the three Chiefs of Staff and the Permanent Under-Secretaries of the Foreign Office and the Treasury, Sir Robert Vansittart and Sir Warren Fisher respectively.

The details of the D.R.C.'s report in February 1934 and the protracted Ministerial discussions which followed have been ably chronicled elsewhere and need not be repeated here.[3] The salient point was that, mainly because of the influence of the civilian members, a policy of conciliation was urged towards Japan, while Germany was seen as the ultimate potential enemy against whom Britain's long-term defence policy must be directed. This was said to be primarily a matter for the Army and the Royal Air Force, whereas in a war with Japan the

Royal Navy would have borne the brunt, at least in the early stages.

The Chiefs of Staff, though unanimous on the need for an Expeditionary Force capable of intervening on the Continent, were extremely modest in their proposals. The Chief of the Imperial General Staff, for example, asserted that the strongest Expeditionary Force the Army could raise would consist of four infantry divisions, a cavalry division and a tank brigade. As Michael Howard has pointed out, such timidity, though pathetic, was entirely understandable. Indeed, Hankey acutely foresaw that the Government would be reluctant to accept even the principle of a Continental commitment for the Army. In the Ministerial discussions of the D.R.C. Report in May and June 1934, Neville Chamberlain, then Chancellor of the Exchequer but already the dominant figure in the Cabinet, led the attack on the excessive cost of the recommendations in general and of the Army's role in particular.[4] He challenged the assumption that Britain had a vital interest in keeping the Low Countries out of the hands of a hostile power who could use them as an air base. Either, he argued, the Belgian frontier defences would hold fast or, if they did not, an Expeditionary Force could not arrive in time anyway. It was far better to give priority to measures of home defence which the public would understand and approve. His idea was to build up the R.A.F. as a deterrent force. These views eventually prevailed in the Cabinet, with the result that the Army's share of the five-year allocation was cut from £40 million to £19 million.

In justifying this modest increase in defence spending in the House of Commons in July 1934 the Prime Minister, Baldwin, made a memorable remark: 'When you think of the defence of England you no longer think of the chalk cliffs of Dover; you think of the Rhine. That is where our frontier lies.' Superficially this sounds like an acceptance of a Continental commitment, but in fact Baldwin was thinking in air rather than land terms. Through the middle years of the 1930s the Chiefs of Staff remained united in the firm opinion that an Expeditionary Force was necessary to maintain the balance of power on the Continent, but for political reasons the Army received the lowest priority in the early, cautious stages of rearmament. The Continental commitment was not yet openly renounced but in practice little was done to prepare an Expeditionary Force capable of fulfilling such a role.

Thus Liddell Hart's R.U.S.I. lecture and his subsequent books[5] emphasizing the concept of a British Way in Warfare appeared at a time when the Army was ill-prepared to send an Expeditionary Force overseas and when public opinion seemed to be strongly opposed to a Continental commitment. A flood of disenchanted books about the First World War in the late 1920s and early 1930s had driven home the lesson: Never Again.

Liddell Hart's thesis was that total commitment to the Western Front in the First World War had been an aberration from Britain's traditional strategy. Britain had become tied to her allies in both policy and strategy as never before. She had committed herself to the fight-to-the-finish formula whose visible symbol was a vast conscript army. The costs of the war of attrition had been excessive:

> Today we are suffering not only from exhaustion of the body, political and economic, but from exhaustion of the spirit. This indeed, has been the gravest symptom and is to be traced all too clearly in our post-war history. The dual cost of the Somme and of Passchendaele, which are so often excused and even acclaimed for their coincident drain on the enemy's man-power, has been deducted from our moral power. There for a generation, if not for ever, has been sunk the faith that created the Empire.

But how had Britain come to adopt a policy so alien to her traditions? Liddell Hart offered this vivid chain of causation:

> The mud of Flanders was symbolical. In past wars we had put our foot in—physically. Before the last war even began we had again put our foot in it—this time metaphorically and mentally. And, during the war, we threw our whole body into it. The immediate chain of causation is to be traced through Sir Henry Wilson's pre-war affiliations, Lord Kitchener's summons to arms, the General Staff's haste to reach France, and General Joffre's haste to reach Germany, down to its ultimate destination in the swamps of Passchendaele. Thither we guided and there we spent the strength of England, pouring it out with whole-hearted abandon on the soil of our allies.
>
> It was heroic, but was it necessary? It was magnificent, but was it war? A supplementary yet separate question is whether it even benefited our allies in the long run. Did we sacrifice our security, our mortgage on the future, for a gesture?[6]

An examination of Britain's major campaigns since the late 16th century led Liddell Hart to the conclusion that 'A romantic habit [such as exaggerating Britain's contribution to the allied victory at Waterloo] has led us to hide, and even hidden from us our essentially businesslike tradition in the conduct of war'. The influence of Sir John Seeley's famous book *The Expansion of England* is evident in Liddell Hart's historical survey, and other naval theorists, particularly Sir Julian Corbett, had earlier advanced a similar maritime strategy, albeit somewhat more cautiously. Indeed the germ of the idea may be traced as far back as Sir Francis Bacon's emphatic declaration: 'He that commands the sea is at great liberty and may take as much or as little of the war as he will.'[7]

According to Liddell Hart, our historic practice was based on economic pressure exercised through sea power:

> This naval body had two arms; one financial, which embraced the subsidising and military provisioning of allies, the other military, which embraced sea-borne expeditions against the enemy's vulnerable extremities. By our practice we safeguarded ourselves where we were weakest, and exerted our strength where the enemy was weakest.

He went on to suggest that the B.E.F. might have been sent to the Belgian coast in 1914, instead of France, where it would have posed a serious threat to the German right wing. But, more important, once the original invasion had been checked, Britain's major military effort should have been shifted elsewhere. The Gallipoli campaign might well have succeeded had 150,000 troops been available at the outset. Further, Britain had erred in adopting conscription: instead she should have given precedence to providing munitions and supplies for her allies. Had such a strategy been adopted, France would have had to refrain from her early futile and costly offensives, while the Central Powers would have been forced to counter the threat in the Balkans, thus weakening their defences in the West. By 1918 the Central Powers would have been closely besieged and economic pressure would have begun to bite all the sooner. Even had a stalemate resulted, leading to a negotiated peace, Britain would have ended up much stronger and with her usual bargaining counters. The conditions of industrial civilization, he concluded, had made our enemy even more susceptible to economic pressure than in the past; and because of our favourable geographical position *vis-à-vis* Germany the Royal Navy was better able to apply it. 'Yet for the first time in our history we made it a subsidiary weapon, and grasped the glittering sword of Continental manufacture.'[8]

Such a summary cannot do full justice to Liddell Hart's colourful language reinforced by a wealth of historical examples. But there can be no doubt about the essential points in his argument or the source of his inspiration. It was the impact of the First World War which shaped his thinking, rather than profound historical study or concern with the immediate defence situation. It is worth emphasizing that the idea that Britain should return to her 'traditional strategy', so avoiding Continental military entanglements, antedated Hitler's rise to power by two years. It was also proclaimed at a time when Britain was pioneering the development of mechanization, whereas her likely opponent was still comparatively weak and forbidden to experiment with tanks on her own soil by the restrictive clauses of the Treaty of Versailles.

In a brilliant reappraisal of the origins of the 'British Way in Warfare' Michael Howard offers a restrained criticism of Liddell Hart's historical arguments:

> It would be doing Liddell Hart an injustice, both as a historian and as a controversialist, to suggest that this analysis of British strategy was anything more than a piece of brilliant political pamphleteering, sharply argued, selectively illustrated, and concerned rather to influence British public opinion and government policy than to illuminate the complexities of the past in any serious or scholarly way.[9]

This is of course correct: the contemporary value of Liddell Hart's preferred strategy was not strictly dependent on the accuracy and validity of its historical precedents. On the other hand there is no evidence to suggest that Liddell Hart did not take his historical data entirely seriously; furthermore, his case derived much of its debating force from the appeal to an allegedly long and successful tradition. In other words it was a vital part of Liddell Hart's case that British strategy in the First World War had been a unique aberration and—as a corollary—

that the traditional maritime strategy was an option still available to her. It is therefore relevant, following Michael Howard and other contemporary scholars,[10] to re-examine the credentials of this traditional, maritime strategy.

Corbett's prolonged study of British naval history led him to a more modest and more persuasive position than Liddell Hart's. In particular, while recognizing the importance of sea power, especially in the *defence* of the British Isles, Corbett perceived that it could seldom, if ever, lead directly to the overthrow of a great Continental Power. The classic example of failure in this sense was Napoleon's triumph at Austerlitz and consequent domination of Europe despite the decisive defeat of the French and Spanish fleets at Trafalgar. In contrast to land victories, naval pressure could only make itself felt slowly, while incidentally causing such annoyance to our own and neutral shipping that strong pressures for a compromise peace were generated.

Furthermore, Britain's interest in the maintenance of a balance of power in Europe and the security of the Low Countries could never be adequately safeguarded by sea power alone. In short, Britain had always required a Continental ally or allies. In fortunate circumstances the latter might be content with a subsidy or a small contingent of British or Hanoverian troops, but in an emergency Britain had had no alternative but to send a sizeable Expeditionary Force to the Continent. In other words the 'maritime strategy' was complementary to or an extension of Continental strategy: they were not alternatives. In British history, as Corbett appreciated, sea power had indeed often been the stronger arm, but the problem was always to harmonize land and sea operations as part of a coherent strategic plan.

The other major weakness in Liddell Hart's historical survey was that he simply assumed the efficacy of the maritime strategy whether implemented in the form of blockade or of amphibious operations designed to assist our allies by distracting part of the enemy's land forces. As regards a war on trade, island powers or powers dependent on the sea for vital commodities (as Spain depended on American silver in the sixteenth century) were in theory highly vulnerable, but in practice it had often proved impossible to exploit this weakness. But great land powers, whether France in the eighteenth century or Germany in the early twentieth century, could at best be embarrassed by loss of their sea-borne trade, not crippled. As for British amphibious operations, with a few exceptions such as the Peninsular War, where Portugal provided a firm base and Napoleon refused to cut his losses by withdrawing, it is hard to find examples which caused the enemy to make significant diversions. Against the few successes, as Michael Howard bluntly put it, British maritime strategy 'had resulted over the centuries in an almost unbroken record of expensive and humiliating failures'. Amongst these he cites Lisbon in 1589, Cadiz in 1595 and again in 1626, Brest in 1696, Toulon in 1707, Lorient in 1746, Rochefort in 1757 and Walcheren in 1809: 'all brilliant in conception, all lamentable in execution'. British amphibious experience in the twentieth century, from the Dardanelles to Norway, Dakar and Dieppe, does not suggest that there has been any marked reduction in the immense 'friction' attendant on such hazardous operations. Furthermore, the advent of railways and aircraft has added to the advantages of the land power in countering such operations. Michael Howard draws two conclusions from his analysis:

> First, a commitment of support to a Continental ally in the nearest available theatre, on the largest scale that contemporary resources could afford, so far from being alien to traditional British strategy, was absolutely central to it. The flexibility provided by sea power certainly made possible other activities as well: colonial conquest, trade war, help to Allies in Central Europe, minor amphibious operations: but these were ancillary to the great decisions by land, and they continued to be so throughout two world wars. Secondly, when we did have recourse to a purely maritime strategy, it was always as a result, not of free choice or atavistic wisdom, but of *force majeure*. It was a strategy of necessity rather than of choice, of survival rather than of victory.[11]

When one turns to consider Liddell Hart's strictures on British strategy before and during the First World War, two general reflections occur. Firstly, he is heavily dependent on hindsight and is therefore unfair to individuals such as Sir Henry Wilson who could not know that decisions taken in 1911 would 'result in' the Somme and Passchendaele campaigns. Secondly, he is extremely simplistic in explaining the shift in British strategy (and the wider issue of the character of the First World War) largely in terms of what he believes to have been the prevalent military philosophy, stemming from the evil ideas of Clausewitz and disseminated by Foch and others in an even cruder and more dangerous form. Without needing to discount such influence entirely, it has to be set in the context of the existing political, social, economic and above all technological environment.

Ivan S. Bloch, the Polish banker, drew the correct deduction from recent experience in his book *Modern Weapons and Modern War* (1900), when he argued that in future the superior strength of the defensive would rule out decisive victories in the Napoleonic style, and that operations would degenerate into a protracted war of attrition. For him the lesson was obvious: war between great industrial powers is no longer viable as an instrument of policy; but it was hardly reasonable to expect the professionals such as Schlieffen to accept so unpalatable a verdict.

It could moreover be argued that the policy adopted towards Germany by the British Foreign Office as well as the War Office in the 1900s was entirely in keeping with traditional strategy designed to preserve a balance of power in Europe. It was simply that from 1815 until the rise of German industrial and naval power Britain had enjoyed a long respite from her traditional anxiety. Even if, as Liddell Hart argued in his lecture, Germany's immediate quarrel in 1914 was with France and Russia, what would happen if she overran the whole of Western Europe and turned France and the Low Countries into satellites? It was very doubtful if sea power alone would then have saved Britain from a humiliating situation.[12]

As for the actual military strategy adopted, though events were to prove the planners mistaken in some details in 1914, it was based on very reasonable assumptions. These were that Germany would go all out for a quick and decisive victory in the West and that the B.E.F., provided it arrived in time, could tip the scales in favour of France. Admittedly, pre-war planners ruled out intervention through Belgium partly on the false assumption that Germany would either respect her neutrality or at least confine her advance to south of the Meuse. But there was also the intractable problem of Belgium's resolute adherence to her neutrality, thus ruling out essential staff planning. Finally, on the outbreak of war the Admiralty refused to guarantee the security of a line of communications to Belgium as distinct from the shorter and more easily defended cross-Channel route.[13]

Liddell Hart was unaware of these considerations, but in answer to a critic of his lecture he conceded that the dispatch of the original B.E.F. was not the irrevocable error; it was rather the total commitment to the Western Front made in 1915. In retrospect this viewpoint has much to recommend it: the cost of attrition between 1915 and 1918 *was* too high. Yet what else was there to do at the time given the impression that France was nearing exhaustion and Russia was on the brink of defeat? Then, as in the 1930s, Liddell Hart and those who thought like him were reluctant to face the unpleasant fact that a hard-pressed ally would not be satisfied by a policy of 'limited liability' on Britain's part. Further, he does not seem to have made the important distinction between the strategic decision and the methods of implementing it. In other words it should be possible to justify the decisions to commit the B.E.F. to France in 1914 and extend the commitment in 1915 without thereby approving the operational strategy and tactics of attrition practised in the last three years of the war.[14]

However shaky its historical foundations, it could still be argued that the British Way in Warfare provided a feasible solution to her defence problems in the 1930s. However, several points may be argued against this contention. In the first place the defeat of Germany in 1918 only briefly obscured the problem of the balance of power in Central Europe. With Russia in the throes of revolution and France permanently weakened by her immense effort during the First World War, it was only a matter of time before Germany's industrial strength made her dominant once again, irrespective of the political colour of her government. Once this occurred, Britain's active participation in Western European affairs, including a definite commitment to send troops, would be more necessary than ever. Foch's celebrated reply when asked by Sir Henry Wilson what would be the smallest force that would be of practical assistance to France, 'A single private soldier; and we would take good care that he was killed', was still unfortunately apposite, though it was not until after Munich that the French again began to ask impatiently for *un effort du sang* from Britain. In sum, the flexible sea-oriented strategy advocated by Liddell Hart 'depended for its success on a continental ally being prepared to accept the sufferings which the British could avoid'.[15]

When one turns from the political to the economic and strategic spheres the drawbacks of the British Way in Warfare are equally apparent. Germany was not critically dependent on her overseas trade; consequently a naval blockade could only be a decisive weapon against her when used in conjunction with a land blockade, and as an additional means of pressure to the constant assaults of armies.[16] In both World Wars Germany was able to counter naval blockade and prolong her war effort against great odds by trading with neutral neighbours, manufacturing a wide range of ersatz goods and, most important, by overrunning vast and valuable areas of eastern and south-eastern Europe. All this of course does not mean that in the long run, particularly in the First World War, the naval blockade did not make an important contribution to the eventual outcome. But what does need to be stressed is that the slogging land campaigns on two fronts sapped the manpower, economy and morale of the Central Powers at a far higher rate than the maritime blockade ever did.[17]

The other side of this coin was that, so far from being able to exploit the advantages of geography and sea power against a vulnerable Continental enemy, Britain, as an island state critically dependent upon sea-borne imports for her very existence, was herself highly vulnerable to naval pressure (in the form of submarines) in both world wars. As Dr Paul Kennedy justly concludes: 'If the Second World War did anything—apart from illustrating the overall decline in the effectiveness of warships alone—it was to break the myth of the efficacy of the "British way of warfare" against a power which straddled half a continent.'[18]

We must now consider another aspect of Liddell Hart's thesis of the British Way in Warfare; namely to what extent it was original and what reception it received in the early 1930s. The first half of the question has already been answered in part: Liddell Hart was original only in his rather extreme and one-sided statement of the maritime case. To go back no further than the 1900s, Corbett, Esher, Hankey and Sir Herbert Richmond had all espoused similar views, and it would hardly be surprising if such a general strategic outlook was popular in naval circles.

The discussion at the Royal United Service Institution following Liddell Hart's lecture on 28 January 1931 provides an interesting variety of criticism and comments. Captain E. Altham, R.N. opened by saying that as a naval officer he was naturally inclined to support the idea of economic pressure, but he queried whether it was wise to generalize the lessons of history. In the First World War economic pressure, as an alternative to more direct and immediate aid to our allies, would have been much too slow in taking effect: France and our lesser allies would have gone under at an early stage, and we should have been left to wage the war alone.

After Admiral Sir Herbert Richmond had expressed himself in general agreement with the lecture, Colonel F. Alston suggested that the French collapse in August 1914 had left the British Government with no alternative to throwing our whole weight willy-nilly into France. Colonel L. P. Evans pointed out that the lecturer had neglected the vital factor of national armies: 'If a country possesses a national army and starts to fight with it, it stakes the bulk of the manhood of the country; if it is beaten it is lost ... These are not the armies with which Marlborough fought.' Fear had impelled us into France: we could not afford the risk of German submarines in the Channel ports.

Air Commodore W. F. MacNeece Foster challenged Liddell Hart's suggestion that in 1915 Britain had a real option of withdrawing from the Western Front and making her main effort elsewhere; surely we had to consider the state of feeling in this country, France and Germany at the time. 'I think the main reason why we were able to win the war was because the French did feel we were absolutely beside them to the last drop of blood, if it came to that, and were not occupied simply in making munitions.'

Major-General J. McM. Walter suggested that if Britain had attempted to enforce her traditional blockade strategy as her major instrument of war in 1914–18 she would have run into very serious trouble with the United States. Major G. I. Thomas made another point which would also be relevant in the late 1930s. In 1914, he stated correctly, France had devised and implemented her offensive plan (Plan XVII) without counting on support from the six divisions of the B.E.F. 'Apparently the problem for the British Government, with an army of six divisions, was to teach strategy to the French with an army of one million.'

Finally the chairman, Lord Ampthill, who had assisted Lord Roberts in his movement in favour of National Service before 1914, took the unusual step of stating that he entirely disagreed with the speaker: Liddell Hart's premisses were wrong and so was his reading of history. 'The fact is that we were doing exactly the same thing as was done by our kings and generals in the days of Agincourt, the Marlborough campaigns, and the Peninsular campaign, but we were doing it with larger numbers of men.'

Two points in the lecturer's reply are worth noting. Liddell Hart conceded Colonel Alston's point that there was really no alternative to sending the B.E.F. to France or Belgium on the outbreak of war, yet he did not modify his scathing criticism of Sir Henry Wilson for doing so when his lecture was later incorporated in a book. Is it too unkind to suggest that to do so would have ruined a memorable metaphor? Secondly, he also accepted the chairman's point that if Britain had had an Army based on National Service before 1914 there was a high probability that Germany would have shrunk from war. Yet he bitterly opposed the adoption of any form of conscription in the 1930s and never changed his opinion that its introduction (in a limited form) in the spring of 1939 had been a colossal blunder.[19]

Liddell Hart's relative lack of receptivity to criticism on points on which he had made up his mind is also evident in his diary note on the day of the lecture:

> Lord Ampthill presided ... He was to have a surprise for I attacked our whole policy of raising a national army and using it in France. Admiral Richmond supported me in an able speech. So did all the sailors, and the only slight criticism came from the soldiers, who had no argument except that of the effect on French feeling. But Ampthill then got up and said he entirely disagreed ... But he made not a single argument, and was content to assert—unhistorically—that the last war was the first of national armies, and that we had not departed from our past history![20]

In contrast to the rather hostile reception of Liddell Hart's original lecture, his correspondence files reveal that many prominent soldiers agreed with him that Britain should never again allow her Army to become entangled in a Continental war. It is noteworthy that these views were expressed in the late 1920s and early 1930s, before the subject had become a most controversial aspect of the rearmament question.

Colonel R. M. Raynsford, the editor of *Fighting Forces*, wrote bluntly as early as 1928 that Liddell Hart should concentrate on pressing for mechanization and dispense with historical lessons: 'We don't want analogies of the past to emphasize the one essential "there bloody well must and shall not be another Somme and Passchendaele in the future".' General Sir Ivor Maxse expressed agreement with the book (*The British Way in Warfare*) as a whole: 'I am much more in favour of the Liddell Hart theory than the Clausewitz theory of warfare. But, my friend, you don't understand politicians. I congratulate you on this. Don't try to.' General Sir John Burnett-Stuart was wholeheartedly in agreement with Liddell Hart's strategic views but chided him for 'trying to make capital out of running down the higher Command'. In 1936 Major-General Fuller, who was then in retirement, wrote that he had lost interest in the Army because there was no intention of modernizing it: 'I fully agree that *in no circumstances* should we use it in a continental war, because, if we do, it will prove nothing short of a suicide club.'[21]

The most penetrating and trenchant critique of *The British Way in Warfare* was, however, written in November 1942, when Liddell Hart unwisely published a new edition of the book in the middle of a total war without revising the contents of the title chapter. Reviewing the book in the *New Statesman*, George Orwell first challenged Liddell Hart's depiction of Clausewitz as the evil genius whose theories had caused the world to plunge into unlimited war in 1914–18:

> There is something unsatisfactory in tracing an historical change to an individual theorist, because a theory does not gain ground unless material conditions favour it.

In Orwell's view, however, Liddell Hart's chief shortcoming was his unwillingness to admit that war had changed its character:

A strategy of 'limited aims' implies that your enemy is very much the same kind of person as yourself; you want to get the better of him, but it is not necessary for your safety to annihilate him or even to interfere with his internal politics.

In 1942 Orwell took the Churchillian standpoint that a compromise peace with Germany was impossible:

Our survival depends on the destruction of the present German political system, which implies the destruction of the German army. It is difficult not to feel that Clausewitz was right in teaching that 'you must concentrate against the main enemy, who must be overthrown first', and that 'the armed forces form the true objective', at least in any war where there is a genuine ideological issue.[22]

It is appropriate to conclude this section with the views of a civilian correspondent who felt that Liddell Hart did not go far enough towards pacifism. Commenting on the proofs of *The British Way in Warfare*, Dr Esmé Wingfield-Stratford, a historian and formerly a Fellow of King's College, Cambridge, argued that it was disastrous for anyone of Liddell Hart's eminence and probable influence to encourage the notion that another war of any magnitude was even conceivable—it would ruin civilization. How could Liddell Hart believe that once war begins 'You can make it a nice, gentlemanly, 18th century affair of pinching bits of territory and holding them for counters? If people were as sane as all that, they wouldn't think of such an insane method of settling their differences. The very forces of mass hatred and mass suggestion that produce war will see to it that war goes on *à outrance* ... This war of the future with military objectives and people winning and all that is our old friend Clausewitz diluted and disguised—it is simply [too] two dimensional for words!'[23]

Wingfield-Stratford had surely touched on a sensitive spot here, for Liddell Hart was too fascinated by military affairs, particularly tactics, to carry his humane feelings through to their logical conclusion. Liddell Hart's close identity with the professional soldier's viewpoint is evident in his note on a talk with General von Blomberg, subsequently Hitler's War Minister, at Geneva in 1932:

A professional soldier, he said he favoured disarmament because it would, by restoring small and handy armies, bring back art,

> leadership, 'gentlemanliness', and the real warrior spirit into warfare. (He lighted up with enthusiasm, truly German, as he spoke of this glorious prospect.)[24]

The Ghost of Napoleon was based on the Lees Knowles lectures given at Trinity College, Cambridge, towards the end of 1932 and published in 1934. In the following year Liddell Hart told a correspondent that though it was almost the smallest of his books, it was the one he valued most—'I think I put more ideas into it than into any other.'[25] In fact the book can most charitably be regarded not as a work of historical scholarship but as a brilliantly written polemic in which Liddell Hart brings to a climax his long-cherished notion that Clausewitz's evil ideas, based on the total warfare of Napoleon's later years, were responsible for the negation of strategy in the First World War.

In contrast to Clausewitz, Liddell Hart praises the more beneficial thinkers, such as de Saxe, Bourcet, Guibert and Jomini, who taught the value of surprise and mobility and could be seen as his own forerunners. T. E. Lawrence commented on a sentence in the draft 'A twentieth century mind was needed to appreciate the full significance of his [Saxe's] proposals': 'This reads as though it had waited for you—quite true, but you don't mean it!' Characteristically, Liddell Hart sides with Jomini against Napoleon's Chief of Staff Berthier, overlooking the former's professional and personal shortcomings. Of Berthier's jealousy towards Jomini he wrote: 'For such men, always, are more anxious to vent their spite, and to spike a possible rival, than to help their country.'[26]

One cannot help feeling that Liddell Hart was prejudiced against Clausewitz in the profoundest sense. He writes throughout as though the latter was *advocating* unlimited war to the exclusion of all alternatives, whereas even a superficial reading shows Clausewitz's intention to have been quite different; namely to suggest that total war was at one end of the spectrum of inter-state violence. Napoleon's campaigns had shown that modern nations in arms were capable of fighting such 'total' wars and, once manifested, it was unlikely that similar wars would not occur in the future. But total war was 'ideal' for Clausewitz only in the philosophical sense: his reiterated phrases to the effect that war must be subordinated to policy and is indeed only 'a continuation of state policy with an admixture of other means' gives the lie to Liddell Hart's misrepresentation. In numerous places where Liddell Hart criticizes him, Clausewitz is only coolly and accurately describing what tends to happen in war. The former gives the game away when he remarks: 'Perhaps the harm might have been avoided if his book had been viewed in the light that its title strictly implied—as a treatise on the nature of war, instead of as a practical guide to the conduct of war.' Yet this is precisely the mistake that Liddell Hart himself repeatedly makes.

The demonstration of the error of the Clausewitzian mode of thinking that was allegedly responsible for the futility of the First World War was not of course Liddell Hart's only aim. In the final chapters 'The Law of Survival' and 'The Liberation of Thought' Liddell Hart called for the formulation of a 'true theory of war':

> It is only commonsense to say that we cannot hope to build up a true doctrine of war except from true lessons, and the lessons cannot be true unless based on true facts, and the facts cannot be true unless we probe them in a purely scientific spirit—an utterly detached determination to get at the truth no matter how it hurts our pride. Not a few military historians have admitted that they feel compelled by position, interest or friendship, to put down less than they know to be true. Once a man surrenders to this tendency the truth begins to slip away like water down a waste-pipe—until those who want to learn how to conduct war in the future are unknowingly bathing their minds in a shallow bath.

He concluded with a stirring appeal for the introduction of a new scientific spirit in pursuit of truth in the Army: 'Loyalty to truth coincides with true loyalty to the Army in compelling a new honesty in examining and facing the facts of history. And a new humility.'[27] What better model for these admirable qualities, one is tempted to ask, than Carl von Clausewitz?

These latter sections of *The Ghost of Napoleon* faithfully reflect Liddell Hart's preoccupation with the pursuit of 'the truth' as reflected in innumerable entries in his private papers in these years. The first half of the 1930s constituted his 'philosophical phase' *par excellence*: the only period in his career when he devoted a great deal of time to writing down abstract ideas and speculations.[28]

Simply stated, these speculations are devoted to two related themes: the need to pursue and publish the truth in a scientific spirit of detachment, and the impossibility of doing either of these things in the British Army. What were the origins of this obsession and why did it come to a head in these years?

The answer is probably to be found in Liddell Hart's passionate involvement, in the later 1920s and early 1930s, in seeking out and publishing the true military history of the First World War,[29] which almost certainly constitutes the largest single subject covered in his enormous archives. Not surprisingly his indefatigable enquiries uncovered an impressive—and depressing—amount of rotten fish: not merely were individual errors and failures glossed over, but important events were completely falsified. What particularly shocked Liddell Hart was the way in which many people in authority displayed indifference or downright hostility to having the record put straight; and also the timidity or false loyalty of those historians who, having discovered the truth, refused to publish it. Chief among the latter culprits, in Liddell Hart's eyes, was Brigadier-General Sir James Edmonds, the leading official historian of the Western Front campaigns, who admitted privately to Liddell Hart that he could not tell the truth frankly in an official history but hoped that it would be evident to those who could 'read between the lines'.[30]

Thus if any one source could be said to have soured Liddell Hart's view of the Army's respect for the truth it was his close contacts with the official historians such as Edmonds, Cyril Falls, Cecil Aspinall-Oglander, and with others 'in the know' such as Sir Maurice Hankey. One may also hazard a guess that the particular philosophizing from which Liddell Hart's reflections took owed something to his close association at this time with T. E. Lawrence, whose biography he was writing and whose intellectual outlook he deeply admired.[31]

A few extracts will suffice to convey a general impression of these extremely interesting reflections carefully recorded by a desperately busy journalist. Liddell Hart believed that Britain's poor record in all her major wars since the Crimea was attributable essentially to lack of clear and honest thinking. 'Unless we are honest about our past,' he noted in 1932, 'and critical about our present, the odds are 100–1 against any improvements in our future—on our next outing.' In a characteristic 'Reflection' the following year he wrote:

> Soldiers as a class have in some ways the highest standards of honour; in other ways the lowest. They are usually the soul of decency—where their class prejudices are not aroused. But it is rare to find a soldier who has a sense of the importance of intellectual honesty, or who puts his duty to the higher cause above his class loyalty ...

In 1934 he went further in writing:

> The *thoughtful* soldier is a contradiction in terms. Custom, convention, and prejudices naturally arising from them, hinder the soldier thinking freely about many subjects, even professions. Scientific habit of thought is impossible. Army never yet ready for any war emergency, nor can it be—suppression of truth ...[32]

The problem of freedom of thought and expression in the Army took on a practical aspect when Liddell Hart began to think about his son Adrian's career. In 1934, when Adrian was only 12, he wrote that although he himself had been very sorry to leave the Army, Adrian would probably not follow in his footsteps because it would entail 'sealing up part of the mind'. When frequently asked if he would like his son to go into the Army he replied 'No, because for all its good points it is no place for a *thinking* man ... and the root of the trouble is the Army's rooted fear of the truth'.[33]

The positive, crusading side of Liddell Hart's zealous pursuit of 'the truth' is well expressed in a letter to Professor Gilbert Murray in 1934:

> To me, increasingly, it seems that the greatest hope of the future lies in encouraging an attitude and fostering a work suggested in one of your earlier paragraphs—in establishing means for the scientific study of war and for the discovery of the truth about it, in its various aspects—its causes and effects, the state of mind in which it is produced and carried on, the effect on its drivers, conductors and passengers. The truth can best be discovered by a scientific study of past wars working hand in hand with the study of psychology.

Another note of the same period could be interpreted as revealing a persecution complex:

> Should stag-hunting be prohibited? It serves as a useful reminder of the way the human pack hunts down the individual of a loftier species. It is the perfect symbol of the conventional Englishman, and, in a worse sense, mankind in general.[34]

Liddell Hart's near obsession with the Army's hostility to 'the truth' was lacking in proportion in that he might have found other hierarchical professions equally impervious and resistant to plain speaking had he been in a position to make an objective comparison. What does need to be reiterated is that his disillusionment was all the more painful for shattering his naïve faith in his seniors when a young subaltern in the First World War. To his credit he recorded this process with clinical detachment in a note in 1935 entitled *Peccavi Contra Veritatem*, where he wrote: 'I am afraid that there has rarely been a young soldier so prone, in uncritical loyalty, to think the best of his superiors, or so eager to credit them with military genius.'

In conclusion, it is clear that Liddell Hart's conception of truth was to some extent limited and subjective. But though conscious of his own considerable ability, he was not lacking a capacity for self-analysis. Thus when a fellow writer on war, Frederick Britten-Austin, referred to him as 'a genius' he noted:

> I'm certainly not conscious of any 'genius'—merely of a moderately smooth-working mind, and a desire to see things clearly. If I have any special power it seems to be that of registering facts that I observe and relating them to each other—my mind works like a cash-register of mental values.[35]

His self-analysis on his fortieth birthday is revealing and worth comparing with those quoted in previous chapters:

> My fortieth birthday—what is my dominant impression? That I'm just beginning to *see*—and, in seeing, to conceive how much lies ahead. Looking back I realise better how much that progress in seeing has been hindered by environment, by pressure of activity, and by moral timidity (a genuine dislike of hurting others' feelings but also a fear of being hurt myself in consequence). I can see myself clearly enough now to realise that those hindrances are still working and I wonder how much stronger they still are than I can see, yet.[36]

The last word in this chapter goes to T. E. Lawrence, whose mocking, humorous, self-deprecatory comments frequently deflated Liddell Hart's idealistic rhetoric in the years preceding his tragic death in May 1935.

In December 1933 Liddell Hart took Lawrence to the Savage Club to discuss the final passages of his biography *T. E. Lawrence: in Arabia and After*. In the Epilogue he proposed to describe Lawrence as 'the Spirit of Freedom come incarnate to a world in fetters', which Lawrence queried on the grounds that he was really an 'anarch' not concerned so much for others' freedom as insisting on his own. Lawrence penned a deflating comment on 'Spirit' and as they came away into the Strand Liddell Hart continued to talk about his desire to get at the root of things. Lawrence, pointing to a match seller, remarked, 'If you let your passion for truth grow upon you like this, you'll finish by selling matches in the Strand'.[37]

Notes

1. Michael Howard *The Continental Commitment* (Temple Smith 1972) pp. 96–98. D. C. Watt *Too Serious a Business* (Temple Smith: U. of California 1975) pp. 88–89.

2. According to a Note on the B.E.F.'s Mobilization Plan in 1932 in the Liddell Hart Papers (11/1932/51) two divisions (1st and 4th) should be ready to embark 14 days after mobilization, a third division would be ready after four months and a fourth after six months. In fact it was doubtful whether the 4th Division would be ready after 14 days and it would be 'certain suicide' to send the 1st and 4th Divisions as at present equipped.

3. M. Howard, op. cit. pp. 105–112. N. H. Gibbs *Grand Strategy* I (H.M.S.O. 1976) pp. 93–268 *passim*.

4. The following exchange in the Committee of Imperial Defence on 22 November 1934 illustrates Ministers' sensitivity to the very notion of an Expeditionary Force:

> Sir B. Eyres-Monsell [First Lord of the Admiralty] drew attention to the phrase 'Expeditionary Force' in a C.I.D. Paper. He asked whether the War Office could see their way to avoid the use of this expression, which, if used in public, would have a bad moral effect. Mr J. H. Thomas [Dominions Secretary] agreed that the expression 'Expeditionary Force' had unpleasant inferences in the public mind.
>
> The Prime Minister [Ramsay MacDonald] agreed, and asked that not only in public, but in all official papers the term 'Expeditionary Force' should not be used. Cab 2/6 266th meeting of C.I.D. discussing C.I.D. Paper 1149–B.

5. *R.U.S.I. Journal* LXXVI 1931 pp. 486–510. *The British Way in Warfare* (1932) pp. 13–41. *When Britain Goes to War* (1935).

6. *R.U.S.I. Journal* 1931 pp. 487–488. *The British Way in Warfare* pp. 14–15.

7. Francis Bacon is cited by Michael Howard in *The British Way in Warfare: A Reappraisal* (24 pp.; Cape 1975).

8. *R.U.S.I. Journal* 1931 pp. 500–503. *The British Way in Warfare* pp. 37–41.

9. M. Howard *The British Way in Warfare* p. 8.

10. See also Correlli Barnett *The Collapse of British Power* (Eyre Methuen: Morrow 1972) Chapter V; Paul M. Kennedy 'Mahan versus Mackinder: Two Interpretations of British Sea Power' in *Militärgeschichtliche Mitteilungen* 2/1974; and N. H. Gibbs' chapter 'British Strategic Doctrine 1918–1939' in M. Howard (ed.) *The Theory and Practice of War* (Cassell 1965: Praeger 1966).

11. M. Howard *The British Way in Warfare* pp. 9–15.

12. Paul M. Kennedy op. cit. pp. 45–52. See also John Gooch *The Plans of War* (Routledge: Wiley 1974) Chapter 6; Brian Bond *The Victorian Army and the Staff College* (Eyre Methuen 1972) Chapter 8.

13. Samuel R. Williamson Jnr *The Politics of Grand Strategy* (Harvard U.P. (1969); N. W. Summerton 'The Development of British Military Planning for a War Against Germany, 1904–1914' (unpublished doctoral dissertation, University of London 1970).

14. M. Howard *The British Way in Warfare* p. 19 and *The Continental Commitment* pp. 54–59. N. H. Gibbs 'British Strategic Doctrine' op. cit. p. 194 makes the point neatly: 'post-war differences of view about how the fighting was actually conducted tended, even if unintentionally, to obscure the basically correct reasons which took Britain into that war in the first place'.

15. M. Howard *The Continental Commitment* pp. 58, 126. George Orwell made a similar criticism of 'The British Way in Warfare': 'But in any case,' he wrote in 1942, '"limited aims" strategy is not likely to be successful unless you are willing to betray your allies whenever it pays you to do so.' See S. Orwell and I. Angus (eds) *The Collected Essays, Journalism and Letters of George Orwell* (Secker & Warburg: Harcourt Brace 1968) II pp. 246–249.

16. P. M. Kennedy, op. cit. p. 51.

17. Ibid.

18. Ibid. p. 63. Dr Kennedy's comments on the declining value of Britain's maritime strategy are made in the context of a persuasive thesis that in the present century sea power itself was waning in importance in comparison with land-based air power.

19. *R.U.S.I. Journal* 1931 pp. 503–510. For Liddell Hart's continuing belief in the folly of introducing conscription in 1939 see his *Memoirs* II p. 235.

20. Liddell Hart Papers 11/1931/1; cf. *Memoirs* I pp. 284–285. For the critical review of *The British Way in Warfare* see *R.U.S.I. Journal* 1932 pp. 680–681.

21. Col. R. M. Raynsford to Liddell Hart 24 September 1928; Sir Ivor Maxse to Liddell Hart 4 September 1932; Burnett-Stuart to Liddell Hart 14 September 1932. See also talk with Burnett-Stuart on 26 August 1935 (11/1935/90). Fuller to Liddell Hart 7 November 1936.

22. *Collected Essays, Journalism and Letters of George Orwell* op. cit. II pp. 246–249. See Liddell Hart's Orwell file for his rejoinder.

23. E. Wingfield-Stratford to Liddell Hart 13 May 1932. In calling the traditional concept of limited war too 'two-dimensional' he was of course referring to the revolutionary implications of air power.

24. 11/1932/9.

25. Liddell Hart to J. Nichols 23 April 1935.

26. For hints that Liddell Hart saw himself in the great line of radical military theorists see *The Ghost of Napoleon* (1933) pp. 66–73, 106–107. *T. E. Lawrence to His Biographers* (1963). *Liddell Hart* p. 134.

27. *The Ghost of Napoleon* pp. 124, 173–185.

28. Some of these reflections were incorporated in Liddell Hart's books in the 1930s but the main collection is *Thoughts on War* (1944).

29. Liddell Hart's books on the First World War include *Reputations* (1928), *The Real War* (1930), *Foch* (1931), *A History of the World War 1914–1918* (1934), *The War in Outline* (1936) and *Through the Fog of War* (1938).

30. Lawrence advised him to 'damn the consequences, and put a footnote, "The official history of our campaigns in France and the East is guilty, here and there, of suppression that amounts to misrepresentation"'. *T. E. Lawrence to His Biographers* op. cit. *Liddell Hart* p. 136.

31. I am indebted to Stephen Brooks for the suggestion of Lawrence's influence. Lawrence told Robert Graves that Liddell Hart 'seems to have no critical sense in my regard': ibid. *Robert Graves* p. 181. Liddell Hart noted that G. B. Shaw's view of truth was many-sided rather than clear: 'He is like a diamond of many facets, that each reflect the light, rather than a great pane of glass like T. E. Shaw [Lawrence].' 11/1933/1 Diary note 22 December 1933.

32. 11/1932/28, 11/1933/28, 11/1934/58.

33. 11/1934/7, 11/1935/37.

34. Liddell Hart to Gilbert Murray 1 November 1934, 11/1934/20.

35. 11/1933/160, 11/1936/2–9 Note dated 21 March 1936. The author recalls Liddell Hart making similar remarks in the last decade of his life.

36. 11/1935/42.

37. *T. E. Lawrence to His Biographers* op. cit. *Liddell Hart* p. 202.

4

Limited Liability, 1935–39

Between 1935 and the outbreak of war in September 1939 Liddell Hart's career reached a peak in terms both of renown and influence. The prestige of *The Times* added weight to his own high personal reputation as a military pundit, while his new appointment as defence correspondent gave him a much wider scope for strategic analysis than he had enjoyed during his decade with *The Daily Telegraph*. Furthermore he had established close and cordial relations with many of the most senior officers in the Army, including successive Chiefs of the Imperial General Staff, Deverell and Gort. Most important of all, he was frequently consulted by Duff Cooper during his period at the War Office (1935–37), and when the latter was succeeded by Hore-Belisha in May 1937 Liddell Hart became his unofficial adviser. During the year of their 'partnership' Liddell Hart was able to exert a continuous and generally beneficial influence in the crucial early stages of rearmament. In this brief period he was able to implement a remarkable number of the reforms which he had advocated for so long: the Army Council was 'purged' and supposedly more progressive officers appointed; the officer career structure was overhauled and streamlined; great improvements were effected in drill, training methods, clothing and living conditions. Particular reforms for which Liddell Hart pressed with varying degrees of success included the development of anti-aircraft defence, a higher priority for the Territorial Army, the co-ordination of defence by a single Minister and the intensification of mechanization of all arms of the service. Above and beyond all these Liddell Hart sought, and as this chapter will try to demonstrate, largely succeeded, to alter the priorities in the roles assigned to the British Army.[1]

Thus, superficially, Liddell Hart appeared to be 'riding the crest of a great wave'; here he was, a mere captain, still in his early forties, with the power of king-maker in military affairs. Yet in fact there was always a darker side to this spectacular success story and in 1939 Liddell Hart's reputation suffered an eclipse from which it would take him many years to recover, and whose effects were never entirely dissipated.

The explanation for this eclipse may be sought at several levels, some of them entirely beyond Liddell Hart's control. The failure of British foreign policy to appease any of her three potential enemies eventually brought about the 'worse case' envisaged by the Chiefs of Staff, to which there was simply no military solution—at least from Britain's own limited resources. In other words Liddell Hart achieved indirect influence late in the day; he was entirely dependent on others to implement his ideas; and his 'patient' had little hopes of recovery in the short time available. More specifically Liddell Hart quickly began to doubt the wisdom of identifying himself so closely with Hore-Belisha:

> I am coming to feel that, from a long-term point of view, the most damaging step I've ever taken was to go in with him. Previous to that I was in an unassailable position, standing apart, yet on good terms with most of the rising generation of soldiers. I put forward my ideas in print and could keep up the pressure in print until they were adopted.
>
> Now every suggestion which I put up, *through H-B*, is resisted. And the people I have helped to put in power are trying to cut off my influence. Worse still, they know who are the men of whom I had a high opinion, and are trying to keep them out. Thus it is becoming dangerous to be, and to be known to be, a friend of mine.[2]

To make matters worse Liddell Hart encountered increasing difficulties in reconciling his own ideas with the *Times* pro-appeasement policy. His articles were frequently held over or distorted by editorial cuts and insertions, with the result that he offered his resignation several times before eventually leaving the paper shortly after the outbreak of war. As if there were not trials enough, his first marriage broke down completely after a long period of strain in 1938, and in the following summer, years of high-pressure work culminated in a heart attack and a collapse due to physical exhaustion.

None of these reasons, however, touches the profoundest cause of Liddell Hart's decline in the estimation of the general public, which was his association with the concepts of limited liability and the superiority of the defensive in warfare. These related themes together provide the keynote in his copious writings in the later 1930s, and even his own skilful justification in his *Memoirs* cannot conceal that, to say the least, there is a paradox to be explained. How, for example, could the outstanding exponent of the theories of armoured warfare and *blitzkrieg* come to be regarded as the champion of the defensive? How could a freelance journalist who prized his independence be viewed as the mouthpiece of the military establishment? How, above all, could it be asserted in a scholarly publication that the doctrine of limited liability 'prepared the road to Dunkirk'?[3]

The aim in this chapter will be to focus on these two central ideas of limited liability and the superiority of the defensive: to explain Liddell Hart's motivation and reasoning; to examine the influence of the concepts and to judge their suitability to Britain's defence needs at the time. It would be foolish to pretend that this can be done in complete clinical detachment; personal opinion must play some part. The task is also complicated by Liddell Hart's approach, prolific output and method of publication. Unlike, say, the Committee of Imperial Defence and the Chiefs of Staff, Liddell Hart was not obliged to survey British and Imperial defence policy as a whole, nor did he have to offer advice on what should be done in particular contingencies. Consequently it is not easy to assess the breadth and comprehensiveness of his thought; he was, for example, certainly less expert in the fields of air and naval policy, and he was more concerned with European affairs than with the Far East. Also, as a brilliant journalist, he was adept at subtly implying the policies he favoured without committing himself too specifically on what should or could be done. He was equally skilful at covering his retreat by the inclusion of qualifying phrases which made it virtually impossible for controversialists to prove him in error. Worst of all, for anyone who attempts to trace the chronological development of his ideas, he is incorrigibly repetitious: articles originally published in newspapers appear again and again in journals and books—sometimes amended but with a great deal of overlapping and even contradiction. His two books *Europe in Arms* (1937) and *The Defence of Britain* (1939) are particularly unsatisfactory in this respect, since both were compiled when he was desperately busy in journalism and public affairs. In the case of the latter he was also too ill to complete a fresh synthesis of his views in the radically new political situation.

As was shown in the last chapter, Liddell Hart's opposition to the commitment of a British Expeditionary Force to the Continent stemmed from his reaction to the First World War and had already been made public in the early 1930s. It therefore pre-dated the specific threat from Hitler's Germany which became increasingly clear from 1933 onwards. It also pre-dated the certainty of Britain's weakness in mechanized forces which, in the mid- and later 1930s, would provide a powerful supporting reason for a policy of limited liability. These points cannot be overemphasized since, it will be argued, Liddell Hart tended—perhaps unconsciously—to cover his most profound objections with more specifically military arguments which seemed extremely persuasive at the time. It should also be stressed that in advocating these ideas Liddell Hart was far from being a lone, heretical voice. Indeed it was precisely because he articulated, albeit with more cogency and military expertise, the heartfelt feelings of a vast number of people in all walks of life—including the Army—that his writings made their immense appeal to public opinion.

Throughout 1935, 1936 and 1937, while the priorities of the Army's roles were endlessly debated by the Chiefs of Staff, the C.I.D. and the Cabinet,[4] Liddell Hart hammered away in *The Times*, in journals, in memoranda and in two books[5] on the theme that Britain should not repeat the fatal mistake of 1914 in committing troops to a Continental war. Despite various qualifications, which will be discussed later in this chapter, the fundamental message inculcated is that the commitment itself was dangerous and should be avoided. Why otherwise would Liddell Hart frequently include his potted history of the 'British Way in Warfare' designed to demonstrate that over five or six centuries the avoidance of mass Continental warfare had always paid dividends? Indeed he noted at the time that his historical studies had predisposed him towards isolation in strategic policy, though technical conditions led him to see the need for collective security.[6] But he believed that Britain would be well advised strictly to limit her contribution to the latter.

Liddell Hart's ideas on the unwisdom of a Continental

commitment are conveniently set out in the chapter entitled 'The Role of the British Army' in *Europe in Arms*. It is a chapter which Neville Chamberlain, shortly after he became Prime Minister, particularly urged Hore-Belisha to read.[7] The following extracts contain the core of the argument:

> When all the conditions are carefully weighed, the balance seems to be heavily against the hope that a British field force on the Continent might have a military effect commensurate with the expense and the risk. I cannot see an adequate prospect, even when the present programme is complete, of it possessing the power of attack necessary to wrest from an invader any ground he may have gained before it could arrive. To fit the picture into the actual frame, formed by the zone between the Meuse and the Channel, serves to deepen this impression. And I do not see that a larger force would have a better effect, nor that subsequent reinforcement might make a great difference, for the limiting conditions have little to do with numbers of men. They are essentially qualitative and technical. Moreover, beyond all the difficulties which face the attacker on land lies the danger of his approach being dislocated by hostile attack from the air. And the larger the force the greater the danger.
>
> It may still be considered that the force is worth while on political grounds. That may be a just opinion. But we ought to be sure that those who decide do so with their eyes opened to the fundamental military limitations ...
>
> If now, there is a doubt as to [the] Belgians' attitude, the difficulties of using a field army effectively may be augmented, and its risks also. If, on the other hand, we should decide to give up the idea of intervention with an army, our military problems would be greatly simplified. We could concentrate on making the Army at home an adequate Imperial Reserve, for the reinforcement of the overseas garrisons, the mobile defence of the overseas territories, and for such expeditions, truly so-called, as may be needed to fulfil our historic strategy under future conditions. We could adjust its scale, organisation, and training to this role – thus avoiding the wide, and increasing, divergence between what is required for offensive warfare on the Continent and what is best suited to mobile defence, to Colonial warfare, and Imperial policing. The forces we would have would need far less development. The very qualities in the average officer which raise doubts of his 'offensive' ability under modern conditions are admirably suited for the performance of these duties. This applies to the men also. Since Marlborough the British Army has rarely shone in the offensive, not through want of

> courage, but from lack of aptitude. It has been superb in defence, and unrivalled as an agent in maintaining order. Those qualities will have enhanced value in proportion as strategic mobility is added.

He concluded strongly that such a policy was best suited to the existing realities of warfare on land. It would not suit those 'who desire the totalitarian training of the nation for war, as well as the preparation of a potentially unlimited army for intervention on the Continent ... The haphazard way in which our part in the Continental land struggle, with its exhausting demands, was determined in August 1914, should be a warning to this generation.'

> Today, owing to technical and tactical conditions, the dispatch of a field army to the Continent does not seem to offer any promise of effect adequate to the risks involved. Since the blankness of the prospect is due to the inherent superiority of modern defence once this has had time to consolidate itself, it is hard to see that in preparing larger forces, to send still later, there is any greater prospect—except of greater waste of lives and resources. On the other hand, with a larger army the risk of it being hamstrung by air attack on its communications would be increased. The advocates of a field force themselves recognize that the difficulties of transporting it and maintaining it overseas today would be much greater than in the past. And there is a further aspect to the problem. The more troops we send oversea the more shipping we have to divert from the essential task of maintaining the supplies to this country, and the more targets we shall offer on the narrow sea passages—which are the most exposed to air and submarine attack. The problem of feeding this country in war is difficult enough without adding to the burden by superfluous and unpromising land campaigns on the Continent.[8]

Thus, though Liddell Hart cannot be accused of ignoring the arguments in favour of a Continental commitment, he put forward the case that Britain should limit her contribution to collective security to an air force contingent and naval support. The idea that a Royal Air Force fighter contingent could replace the Army's traditional role in an expeditionary force found a prominent supporter in General Sir John ('Jock') Burnett-Stuart, who took over Southern Command in 1934.

Burnett-Stuart also caused great irritation in the General Staff by his trenchant opposition to a Continental commitment of the Army in letters to *The Times*.[9]

Liddell Hart's objections to a land Continental commitment can be summarized as follows:

1. Britain cannot afford to rearm all three services simultaneously with the same priorities in expenditure: the air force and the navy will give better value for money.
2. The Royal Air Force can take the Army's place in providing a Continental contingent.
3. The defensive is immensely superior in land warfare and Britain lacks mechanized forces suitable for offensive operations.
4. A Continental ally [i.e. France] does not need our small infantry force but would prefer mechanized divisions.
5. It is extremely doubtful if a British expeditionary force could arrive in time to affect the outcome.
6. It is no longer essential to defend Continental bases [i.e. in Belgium] with a land force because of the increasing range of aircraft.
7. Even if Britain does produce mechanized forces they will be of more value for the active defence of our overseas territories.
8. The British high command has not learnt the lessons of the First World War and indeed is preparing to repeat the mistakes that led to the Somme and Passchendaele. Hence even a token commitment is to be avoided since it would probably lead to the sending of a mass army.
9. The British temperament is unsuited to offensive operations.[10]

Clearly not all of these points were of equal validity and weight, but Liddell Hart's technique was to pile argument upon argument to achieve a cumulative effect. The evidence in his files shows that those already of like mind were delighted by his articles, but doubters and critics were apt to find his style irritating and his argument unconvincing. Colonel Pakenham-Walsh's verdict on completing *Europe in Arms* may be cited as a typical 'middle of the road' reaction:

> It is really good and gives a great deal of food for thought. But LH gets pet catch words and ideas and drives them mad, quickly getting away from the fundamental points and applying the idea *in extremis* ... his own examination of problems is not really scientific, but is swayed too much by his catchwords.[11]

To examine the validity of Liddell Hart's supporting arguments one by one would entail writing a history of the evolution of British defence policy in the 1930s. Suffice it to say that from the standpoint of British national interests and on a purely military level there was much to be said for a policy of limited liability. In the mid-1930s the Army was indeed pathetically unprepared to participate in a European war; even the leading regular infantry divisions (there were only five) were lacking in vital weapons, stores and specialist personnel, while not a single armoured division yet existed. Moreover, so far from the small available forces at home being concentrated and organized for a European emergency, the Abyssinian crisis led to a significant and—as it proved—permanent reinforcement of the Middle East.

Again, as regards mechanization and progressive leadership, Liddell Hart's *Memoirs* chronicle in depressing detail the sad falling-away after a hopeful period of experiment in the early 1930s and the side-tracking of the outstanding exponents of armoured warfare such as Hobart, Martel, Pile and Lindsay. Liddell Hart could also be excused for thinking that the General Staff were bent on resurrecting a B.E.F. consisting essentially of infantry divisions, only weaker than its predecessor of 1914 in numbers, training and artillery support. We can also now see from the records that the General Staff was extremely confused about the B.E.F.'s role once it arrived on the Continent.[12] He was, furthermore, correct in his most important qualification that if there had to be a Continental commitment on land it should take the form of mechanized divisions—even one or two would be more useful to the French in the counter-offensive role than twice as many slow-moving infantry divisions.

Some of his other arguments were questionable because they depended on assumptions about the timing and balance of forces at the outbreak of a war. In view of the extremely fluid diplomatic situation throughout the 1930s Liddell Hart tended to be too dogmatic and general in his assertions. Belgium's return to a policy of neutrality in 1936, for example, greatly increased the difficulty of sending a small B.E.F. thither to protect advanced air bases, while the ever-increasing range of aircraft somewhat reduced the importance of such bases. It was far from clear, however, that these developments rendered Britain's traditional strategic interest in the Low Countries invalid. An equally difficult problem, which gave the Chiefs of Staff endless trouble, was whether the B.E.F. could be safely transported across the Channel and arrive in the theatre of operations in time to affect the outcome. On the whole they decided that it *could*, even though obliged to operate through France's western ports to avoid the worst of anticipated enemy air attacks—and of course they proved correct.[13] In a rare admission of error Liddell Hart remarked in his *Memoirs* that in pressing for the need for very early intervention by any expeditionary force sent to the Continent, he did not foresee that if the French were not directly attacked themselves they would be content to mark time while the German mechanized forces overran their eastern allies, instead of taking the offensive themselves in aid of those allies.[14]

In general, then, we can conclude that Liddell Hart was on strong ground in focusing his criticism of the Continental commitment on Britain's *military* unpreparedness and inability to play a significant part on land in deciding the outcome of a great European war. It is noteworthy that Liddell Hart's approach to this complex issue was first and foremost that of the tactical, operational expert. His starting point was the likely influence of new weapons and technology on tactics and on this basis he constructed his theory of the probable course of strategic operations. Strategic problems then led him to consider foreign policy, but this aspect receives comparatively little emphasis. Such an approach had its merits; after all, the inter-war period witnessed the burgeoning of numerous policies for which the necessary forces did not exist, and it was useful if this was occasionally pointed out. There was, however, a real danger of the tail wagging the dog, i.e. of tactical considerations being allowed to determine military and hence foreign policy. Before discussing the practical outcome of Liddell Hart's limited liability concept it is essential to examine the underlying belief that technological innovation was rendering the defence ever stronger in comparison with the attack.

A major theme in Liddell Hart's publications in the mid- and late 1930s is that the defence is markedly superior to the attack in modern land warfare and that weapon developments are actually increasing this superiority. This thesis was advanced, for example, in three curious articles under the general heading 'Defence or Attack?' in *The Times* on 25, 26 and 27 October 1937. In the first article he was on strong ground in criticizing the slow development of mechanized forces and new tactics in the British Army, but in the second and third articles he repeatedly suggested that the initial attacker had rarely succeeded in any battle in the past six centuries and had even less chance of doing so at present.

> Modern weapon development has been predominantly defensive. It was the machine gun which, above all, established the superiority of the defensive in the last War; and today there are more machine guns than ever. The anti-tank and the anti-aircraft gun, weapons which have seen the most improvement since the war, are purely defensive.

Already in *Europe in Arms* he had disputed the view that mechanized divisions would be able to pierce the defences in the early days of a war except where the enemy was taken by surprise and himself unmechanized. By using railways, rivers and canals as barriers, and by demolishing bridges and blocking defiles, the defender could go far to nullify the new menace. Moreover mechanization could enable the means of obstruction and demolition to be moved more swiftly to any threatened spot. 'Despite the apparent advantage that mechanization has brought to the offensive, its reinforcement of the defensive may prove greater still.' Nor did he believe that the development of air power could radically alter the balance.[15]

The lesson was clear and encouraging for non-aggressive nations: 'So great is the power of the defensive nowadays that a small reinforcement may suffice to establish a deadlock.' Victims of aggression were unlikely to be beaten provided they refrained from 'foolish indulgence in attacks'. Developments in the present century gave cause for doubt 'whether any form of offensive action remains essential to the purpose of a non-aggressive State in war'. In a concluding paragraph much more appropriate to the nuclear age than to the 1930s, Liddell Hart speculated that a combination of the indecisiveness of war and the mutual fear of air reprisals would rule out unlimited war between Great Powers. 'In that case war *à outrance* would disappear and be replaced by a reversion to the eighteenth-century game of "playing for points".'

In all his discussions of the relative strengths of the attack and the defence it is evident that Liddell Hart was chiefly worried that the British and French High Commands were bent on repeating the futile offensives of the First World War. Indeed the C.I.G.S., Sir Cyril Deverell, vainly tried to persuade him that his (Deverell's) remarks after a war game had been misunderstood and the British Army was not in fact wedded to a tactical or strategic doctrine of attack in all circumstances.[16] Even more surprising, in view of their construction of the Maginot Line, heavy reliance on a short-service conscript army, and timid reaction over the German reoccupation of the Rhineland—not to mention the profound impact of their losses in the First World War—was Liddell Hart's attribution of an offensive mentality to the French. Ironically, it would probably have stood the Allies in good stead if the French had launched a full-scale offensive on the Western Front in 1939.

In the controversy arising from these articles and other interchanges in the service journals,[17] the sweeping assertions and absence of precise definitions creates the impression that the 'attack versus defence' question is about as sensible as asking 'how long is a piece of string?'. Although the disputants ranged far and wide through history (from Crécy to the recent campaigns in China, Abyssinia and Spain), it ought to have been clear that circumstances alter cases and that the unexpected frequently happens in war. What does emerge is that Liddell Hart did *not* then foresee the *blitzkrieg* offensive operations which Germany and Japan carried out so successfully between 1939 and 1941.

The grounds on which Liddell Hart's thesis was most frequently criticized were whether a European war could be strictly limited, and if it could not, would moral and political considerations permit Britain to limit her contribution to an alliance. In retrospect there is an element of tragic irony in the fact that the Chiefs of Staff, and even more the General Staff, were sound in their assumptions that Britain still had vital interests in Western Europe which could not be adequately insured by a policy of 'limited liability' but were conservative as regards mechanization and vague about what the expeditionary force would do after arriving in France. Liddell Hart, by contrast, had progressive ideas on the need for mechanization and the kind of mobile operations to which it could lead, but tended to deny the need for a Continental commitment which alone could have justified a higher priority for the Army in defence expenditure. Where Liddell Hart was unjust was in his allegations that the General Staff had no other idea than resurrecting a mass army on the lines of the First World War. After all the maximum force which could have been raised in peacetime (i.e. without conscription and an all-out war effort) was five Regular divisions and twelve Territorial divisions, and even then several of the latter were earmarked for anti-aircraft defence at home.

The orthodox soldier's viewpoint was probably best summed up by Colonel Henry Pownall's diary entry on 27 January 1936 following a meeting of the Defence Policy Requirements Committee:

> It is obvious that in a Great War the Regulars have to be backed up, and equally obvious there is nobody to do it except the T.A. Those are the facts. Why not face them boldly? But they funked—badly.
>
> There was a further and most dangerous heresy—the Chancellor's. That of 'limited liability' in a war. They cannot or will not realise that if war with Germany comes again (whether by Collective Security, Locarno or any other way) we shall again be fighting for our lives. Our effort must be the maximum, by land, sea and air. We cannot say our contribution is 'so and so' and no more, because we cannot lose the war without extinction of the Empire. The idea of the 'half-hearted' war is the most pernicious and dangerous in the world. It will be 100 per cent—and even then we may well lose it. We shall certainly lose it if we don't go 100 per cent. In God's name let us recognise that from the outset—and by that I mean now. The Chancellor's cold hard calculating semi-detached attitude was terrible to listen to ...

Pownall also wrote on 14 March 1938:

> It is discernible that the idea of a possible continental commitment for the Army is beginning to come back into people's minds. Gort is strong on it. Although it is fourth, and last, in priority for the provision of money, it may come first in priority when the emergency arises. And since it is financially last it means that troops we send will be ill-provided for in their task. The General Staff are constantly accused, by L-H and the politicians, of hankering after sending the Army to France and repeating the 'horrors of Passchendaele.' We don't want it but we believe that it is a highly probable role and we need to be ready for it.[18]

General Sir Frederick Maurice, replying in *The Times* on 29 October 1937 to the 'Defence or Attack?' articles, correctly predicted that 'we will have to send an army to our ally and we will not be free agents in choosing our tactics'. Rather surprisingly, no one on the General Staff, not even Lord Gort the C.I.G.S., seems to have faced up to the fact—until the summer of 1939—that the smaller Britain's land contingent the less say she would have in deciding allied strategy. Only after Munich did the British Government awake to the harsh reality that the French would expect *un effort du sang* as evidence of Britain's full commitment to an alliance. Yet in challenging Liddell Hart's doctrine of limited liability in the *Army Quarterly* earlier in 1938 a French general, Baratier, had protested against the English tendency 'to throw the main weight of the war upon her allies. It cannot be refuted too strongly. If ever Germany attacks our country, she will wage a totalitarian war, in the course of which France and England, in order not to succumb, will have to throw into the balance all their resources on land, on sea and in the air.'[19] These statements, however unpalatable, were closer to reality than Liddell Hart's optimistic view, summed up in a newspaper headline as late as 9 July 1939: 'The Knock-out Blow is Just a Dream. We could lose a war only by trying to win.'[20]

It might be assumed from his near-obsession with limited liability that Liddell Hart would favour the policy of appeasement towards Britain's potential enemies in the 1930s, but this was not so. Rather, he supported the principle of collective security with Britain making a contribution in a form suited to her strategic conditions and resources. Over Abyssinia in 1935 he vainly attempted to persuade Geoffrey Dawson, editor of *The Times*, that the paper should press for full economic sanctions and shutting off oil supplies to Italy, even though the almost certain outcome would be a war in the Mediterranean.[21]

He does not appear to have taken an equally firm line over Germany's reoccupation of the demilitarized zone of the Rhineland in March 1936, but to have viewed the matter in political and ethical rather than strategic terms. He shared the widespread British reaction that 'Laying aside the breach of faith, the right thing has been done in a wrong way'. In a discussion shortly afterwards with Colonel Bernard Paget and Brigadier Sir Ronald Adam, Liddell Hart suggested, according to his own note, that if France attacked Germany the latter might launch a counter-offensive through the Ardennes. The soldiers made the point that the Belgians would not feel happy until a British field force arrived on their soil, and both believed that such a force could arrive in time because Germany would not be ready in the near future to take the offensive in the West.[22]

Another issue on which Liddell Hart did believe Britain should make a stand was over the Axis Powers' supply of arms and troops to General Franco's side in the Spanish Civil War. Indeed in a memorandum he wrote for Hore-Belisha in March 1938, just after the German take-over in Austria, Liddell Hart argued that the 'military key' to the rivalry between Italy and Germany and Britain and France lay in Spain:

> As I have been pointing out for eighteen months a German-Italian domination of Spain would place heavy odds against the success of Britain and France in a war with these powers. Any one must be blind who cannot see that victory now for Franco spells this domination. Militarily it has been much easier for us to prevent that victory than for Germany and Italy to secure it. The cards are still in our hands—until Franco and his allies secure the eastern seaboard of Spain. When that happens the whole game is likely to be lost.

His solution was for Britain and France to drop the farce of non-intervention and supply the Government side with enough material resources to restore the balance: if a competition in supply of aid should ensue, Britain and France enjoyed the geographic advantage.

> If they dared to press their objections to the point of war, we should

> fight with all the advantages of the defensive and under more favourable circumstances of strategic geography than we could hope for once Spain has been conquered. For these reasons the risk seems less than that presented by any other contingency that can be foreseen.[23]

The outcome of course was that the Allies did not interfere; Franco triumphed with significant aid from the Axis but—contrary to Liddell Hart's reasonable fears—did not intervene to cut off Britain's supply route through the Mediterranean in the Second World War.

Liddell Hart's disagreement with the *Times* editorial pro-appeasement policy reached a climax over the Munich crisis. Since he had made his position clear at the time of the first Czechoslovakian crisis in May 1938, neither Dawson nor Barrington-Ward consulted him before publicly advocating that the Sudetenland, with Czechoslovakia's fortified zone, should be ceded to Germany. Liddell Hart clearly perceived the fatal strategic implications of such a sacrifice and later wrote that he was shocked and nauseated by the tone of the *Times* leaders. Nevertheless his appreciation of the situation to a large extent echoed the pessimistic assessments of the Chiefs of Staff. True, he doubted whether the Germans were capable of conquering the whole of Czechoslovakia so long as France provided a distraction in the west. On the other hand he did not expect France to carry out a serious offensive either on land or in the air. He concluded rather lamely that 'The best hope of compelling Germany to relinquish her lodgment [in the frontier zone of Czechoslovakia] would be the development of general economic pressure, which would be reinforced by moral isolation'. The main lesson he derived from the crisis was that Britain and France were inhibited from taking offensive action by the weakness of their defences against air attack.[24]

How did the outcome of the Munich crisis affect Liddell Hart's views on limited liability? Having pressed successfully for the preparations for a land Continental commitment to be relegated to the Army's lowest priority, within a few months Liddell Hart began to doubt the wisdom of the decision. As early as 17 June 1938 he suggested in a *Times* leading article that the increased tension in Europe might necessitate the dispatch of a field force to the Continent, though he stressed that a mechanized force would be more valuable to the French than infantry divisions. In July both Paul Reynaud and the French Military Attaché, General Lelong, urged that Britain should be prepared to send mechanized troops—the former suggested the modest total of 60,000. Liddell Hart was to argue tentatively in 1939, and more firmly in later years, that the shattering of France's alliance system, together with the loss of the Czechoslovak forces and arms factories to Germany, had transformed the situation and had convinced him that Britain must after all endeavour to restore the balance by accepting a Continental commitment on land. Critics of the limited liability policy would argue that the Munich settlement did not so much create a radically new situation as underline the necessity of Britain playing an active part in the defence of her vital interests in Western Europe. The reaction of Major-General Sir Henry Pownall, now Director of Military Operations, is a case in point:

> Liddell Hart produced two articles in *The Times*. The first [17 June] on the employment of the Field Force. He shows anxiety that recent events have made the needle swing towards the Western commitment, which he dislikes. But it is interesting to note that after confused writing and his usual false deductions from History he concludes that 'our infantry divisions ... might wisely be reserved for other contingencies to which they are more suited and for which it is essential that we should have forces available ... Belgium ... Holland ...'. So his needle is swinging too![25]

Liddell Hart's writings in the months following Munich do not suggest that his conversion to the need for a Continental commitment was either sudden or complete. In fairness it should be recalled that he was by now *persona non grata* with Dawson and Barrington-Ward and found it increasingly difficult to get his views on the deteriorating strategic situation published at all. For example, two important articles entitled 'An Army across the Channel?' which he wrote on 7 December 1938 were held over until 7 and 8 February 1939, when they were published in the wake of Chamberlain's announcement that he had assured the French of Britain's support on land.

Even so, without Liddell Hart's later explanation in his *Memoirs* available to them, it is easy to understand why contemporary readers (and the compiler of *The History of The Times* later) should gather that his intention was still to *oppose* the Continental commitment. Major-General Sir Edward Spears, for example, wrote that he had read Liddell Hart's article (on 7 February) 'with interest and at the same time with great concern. . . . I don't know what is coming, but I beg you not to lose sight of the fact that from the recent very comprehensive examination of French opinion which I carried out, it was quite clear that France will not consent to do all the fighting whilst we, under any pretext we may choose to produce, only prepare for the kind of warfare which will not involve similarity of sacrifice . . .'. Liddell Hart's explanation is that he had to recapitulate all the points on which he and Barrington-Ward were in agreement before slipping in the bitter pill at the end. Certainly he had already advised Anthony Eden in January 1939 that as a result of Munich, British land reinforcements to France were now necessary in the event of war. Where he did his own reputation considerable harm was in re-publishing 'limited liability' arguments in *The Defence of Britain* in July 1939, by when Britain had entered into a more definite military commitment to France than before 1914 and when it was clear that the B.E.F. would cross the Channel on the outbreak of war whether mechanized or not. While accepting that Britain could not afford to run the risk of seeing France overthrown, he still questioned the need for a land contribution in her support.

> When account is taken of the power of modern defence, the limited length of the Franco-German frontier in relation to the size of the French Army, and the strength of the fortifications there, it is not easy to imagine that any assault upon it could have much chance of success. The most chance would lie in an initial surprise; but our field force could hardly be on the scene in time to help in parrying this. Once the French Army had mobilized, that field force, in which infantry predominate, would add but a trifle to their strength. Its value to them might be much less than the risk to us of being drawn, by degrees, into a fresh mass-effort on land that would offer less prospect of result and more of exhaustion than even in the last war.[26]

What is one to conclude in the strange case of Liddell Hart and the doctrine of limited liability? Although the arguments for a small, high-quality mechanized force—which Britain did not possess before 1939—sound plausible, there is a strong impression that Liddell Hart was fundamentally opposed to the idea of sending even a single soldier to the Continent lest the dreadful experience of 1914–18 should be repeated. Such an attitude is of course very understandable and it was widely shared. But whether Britain ever had an option in the matter may be doubted unless the Government was prepared to push appeasement even further by seeking a non-aggression pact with Hitler. After March 1939 even that unlikely escape route disappeared.

Had Liddell Hart ever been fully converted to the need for a Continental role for the Army, would his influence have made any difference? The influence of writers and outsiders generally on military and political decisions is notoriously difficult to evaluate. Between 1935 and 1939 Liddell Hart's ideas appeared in the influential columns of *The Times*, whose editor and his deputy are known to have enjoyed very close relationships with many political leaders, particularly in the Conservative Party. This at least went far to ensure that he would be read in the right quarters.

In general, except where an outsider is in regular contact with a Minister—as Liddell Hart was with Hore-Belisha in 1937–38—his ideas are more likely to clarify and buttress pre-existing trends of thought than to initiate radically new ones. In the case under consideration, Neville Chamberlain and many others were instinctively reluctant to accept a Continental commitment from the moment the subject became a major defence issue following the Defence Requirements Committee's first report in February 1934. As the controversy dragged on, however, it was doubtless valuable to have the public support and detailed arguments of *The Times*' military correspondent, whose name (though officially anonymous) was widely known.

When one turns from the abstract problem of the effect of thought on action to such practical questions as whom did Liddell Hart advise, what did he tell them and how did they respond, then there is ample evidence that he exerted a very considerable influence indeed. True, most of the evidence, brilliantly marshalled and presented in his *Memoirs*, comes from his own enormous archives; so occasionally he may unwittingly misrepresent another's view or, conversely, misjudge the impact of his own opinions. But against this, his passion for getting at the truth must not be underestimated. Also, his lifelong habit of circulating innumerable copies of his memoranda provided a useful check on errors of fact or opinion.

It is ironic to reflect that he probably exerted more influence in Britain on the shaping of policy and strategy through his doctrine of limited liability than on the development of armoured forces, with which his name is often more linked.

To take Liddell Hart's relations with the soldiers first, it is evident that many in the senior ranks shares his opposition to a Continental role for the Army. This attitude may be attributed to two related factors: their experience in the First World War and their first-hand knowledge of the Army's complete unpreparedness to face a major European Power. General Sir John Burnett-Stuart, for example (whose name was usually linked with Liddell Hart's as the leading opponents of a European role for the Army), wrote to *The Times* as late as March 1939:

> Academic studies as to whether we should or should not send an army across the Channel to support France if she is attacked are interesting. But the immediate question whether a British Army would actually be sent ought to be an easy one to answer. We have no Army to send to the immediate assistance of France; nor, unless we are prepared to add largely to our Regular establishments and to find a much more effective way of filling them with men and trained reservists than exists at present, shall we ever have one.

Similarly, on 5 February 1938 General Sir Edmund Ironside noted in his diary:

> We have no Continental commitment now. I told the Commanders' Conference that our wretched little Corps of two divisions and a mobile division was unthinkable as a contribution to an Army in France. Nothing behind it either. Let us make Imperial plans only. After all, the politicians will be hard put to refuse to help France and Belgium when the 1914 show begins again.

This last sentence is revealing, for although many other officers expressed 'anti-Continental' views, including Fuller, Pile, Gort and Montgomery, most of them realized that whatever priorities were assigned to the Army in peacetime, if war again broke out in Europe a British force would have to be sent, however ill-prepared.[27]

Few generals seem to have grasped this unpleasant fact more firmly than Sir Henry Pownall, whose main task on becoming Director of Military Operations in January 1938 was to persuade the Government to accept that the Army must be prepared for a European role. This largely accounts for his bitter remarks directed at Liddell Hart, Burnett-Stuart and others who provided comforting arguments for the opposition. In his first interview with Hore-Belisha on becoming D.M.O. Pownall bluntly told the Minister:

> It was my firm belief that if we got embroiled in a major European war sooner or later, and I believed sooner, British troops would go to France. If that was not recognized or [was] glossed over in peace then the troops would go untrained and ill-equipped for the purpose—with dire results.

In a Note in February 1938 deploring the success of the Minister for the Co-ordination of Defence (Sir Thomas Inskip) in relegating the European role to the Army's lowest priority, Pownall concluded:

> My view is that support of France *is* home Defence—if France crumbles we fall. Therein lies the fallacy of his argument. And see how the role is altered to fit the purse. The tail wagging the dog. One can see he doesn't like it for he asks his colleagues to 'share his responsibility.' If France were to be overrun, he says, and we had to send an Army to help, the Government of the day 'would be criticised for having neglected to provide against so obvious a contingency.'
>
> A sorry story. What about the unfortunate troops?[28]

Contrary to Liddell Hart's implications, it is difficult to find a single responsible soldier who believed that it was either possible or desirable to raise a mass army, mainly composed of infantry, on the lines of 1914–18. Nor did they advocate conscription, because in fact they were in agreement with Liddell Hart that it would further handicap the training and equipping of existing forces. The weakness of the typical General Staff officer, such as Pownall, Haining and Dill, was that they were lukewarm towards mechanization and vague in their ideas as to precisely what Continental role a small under-equipped B.E.F. could play. On the other hand the Government's delay in not definitely deciding until February 1939 that the Regular Army (not the Territorials even then) must be prepared for a European war, had very serious effects on every aspect of organization, planning and equipment. What would have happened to the British Expeditionary Force had Hitler launched his intended attack in the West in the autumn of 1939 hardly bears thinking about.

Duff Cooper, one of the few Cabinet ministers to support the Continental commitment, told Liddell Hart that he was very sorry to leave the War Office (in May 1937). According to Liddell Hart's note on their conversation: 'He had argued with Neville [Chamberlain]. But N influenced by mine [i.e. LH's] and Jock Stuart's arguments about B.E.F. and felt somebody else better to implement his policy. Unfair to D.C. D.C. demurred and said he hoped to carry on, and could carry out [the policy], but Neville adamant.' The C.I.G.S., Sir Cyril Deverell, told Liddell Hart ('in a friendly way') that his articles in *The Times* had increased his difficulties a hundred per cent in getting money from the Government for the Army. They led the ministers to argue that 'only policemen were needed'. Ironside had the same impression, telling his officers that 'the C.I.G.S. is right up against it trying to get money for things. Neville Chamberlain won't give money for the army as some of Liddell Hart's and other recent articles might appear to mean we do not need an army.' A fellow enthusiast for mechanization, Brigadier Martel, expressed anxiety to Liddell Hart that Burnett-Stuart's letters to *The Times* would impede the development of tanks.[29]

These suspicions are borne out by the documents now available in the Public Record Office. For example, Liddell Hart's letter in *The Times* on 11 November 1936 on 'The Role of the Army' provided the starting point for a bulky file of memoranda and letters in which Pownall, Hankey and others attempted to convince Inskip, Hore-Belisha and the Cabinet of the need for a Continental role for the Army. One of the Government's main aims in abandoning such a role in February 1938 was to reduce expenditure. Lord Gort informed Hankey that 'His [Inskip's] big saving of £14 million had been made largely on tanks and partly on ammunition. He had dropped the idea of heavy and medium tanks and fallen back on cruiser and light tanks.' Gort pointed out that these cuts would mean that the Army would not be adequately equipped even for a war against Italy in Egypt.[30]

Liddell Hart's preoccupation with the superiority of the defensive, coupled with his belief that it should not be too difficult to deter an aggressor, stood him in good stead on the subject of the air defence of Great Britain (A.D.G.B.). By the late 1930s he had given up his brief flirtation with strategic bombing and was increasingly confident that defensive measures could be effective against it. Although he had fewer contacts with prominent airmen than with soldiers (and fewer still with sailors) he did enjoy close relations with successive Permanent Secretaries at the Air Ministry, Sir Christopher Bullock and Sir Donald Banks. Throughout the 1930s, and particularly as a consequence of the weaknesses revealed during the Munich crisis, Liddell Hart championed the cause of home defence against widespread pessimism based on the belief that 'the bomber will always get through'—and cause enormous casualties. In December 1936 Air Marshal Joubert de la Ferté (Commander, Coastal Area) estimated that Germany could launch a surprise air attack on Britain within two hours of war being declared. We should do well to bring down one bomber in nine. The enemy could only keep up continuous attacks for a month, but would the country stand a month's hammering? Sir Donald Banks was, if possible, even more of an alarmist. Liddell Hart noted after lunching with him in May 1937:

> Seriously concerned with tonnage of bombs which could now be dropped on England in a day—2,000 aircraft, over half bombers with load of a ton, or more.
>
> Thinks Germans have no hint of our plans for an alternative capital, just as we are unsure of theirs. Buildings observed at Munich may be only a blind, and they may intend to stay in Berlin, or suburbs.
>
> It may well be necessary to regard London as untenable.
>
> May have to cope with a million causalties—no adequate hospital arrangements, or resources. Problem not yet measured.[31]

Liddell Hart, by contrast, drew some comfort from the lessons of the Spanish Civil War, writing to Ironside:

> In regard to the menace of the Air, for example, we have had to

> depend mainly on deductions from the evidence of technical development until recently. Now Abyssinia and Spain have given us ground for some encouragement as to the deterrent effect of anti-aircraft fire, and the actual effect of fighters, so one can know that, if still cautiously—since the number of bombing machines in Spain has been extremely small, even by the standards of the last war.
>
> I find, too, that despite this encouraging evidence the anti-aircraft experts here still tend, in private confession, to be very dubious of their chances of hitting any bombers, especially under our climatic conditions. On the other hand, I think we are on more solid ground in putting a modest estimate on the material, as distinct from the psychological, effect of bombing. And even the psychological effect may be minimised by proper preparation and education.[32]

His efforts on behalf of air defence were handicapped by his ignorance, until 1939, of the secret development of radar. This had one amusing sequel. In May 1937 he sent Sir Thomas Inskip a paper on 'The Protective Power of Credible Rumours' in which the basic assumption was that 'the danger we most have to fear is that of air attack on our cities and ports'. Liddell Hart suggested that a rumour be generated that we had discovered a new antidote to air attack, keeping the secret of its non-existence to the smallest circle. In November Air Marshal Sir Wilfred Freeman and Group Captain Jack Slessor (who clearly did know about radar) unwittingly told Liddell Hart, the rumour's source, that they were anxious about Inskip spreading the idea that we might have a death ray or similar device as a defence against air attack.[33] Here was an instance where the outsider's attempts to help could have counter-productive results.

Liddell Hart's detailed account of his *Memoirs* shows that, particularly during his partnership with Hore-Belisha, he was able to exert considerable influence in drawing attention to the need for much greater preparedness of ground and fighter defences against air attack. Here again, however, he made enemies, both in the members of the Air Staff who still wanted priority for bombers, and in those of the General Staff who argued (correctly) that A.D.G.B. was receiving priority at the expense of the preparation of the Field Force. Gort, Brooke and Pownall were among Liddell Hart's opponents on what they regarded as disproportionate expenditure on air defence.[34]

Liddell Hart's influence on Government policy through his close contact with individual politicians has been meticulously recorded in his *Memoirs*. The politicians who sought his advice included Eden, Churchill, Hoare, Duff Cooper, Halifax and Lothian.[35] As clearly suggested above (p. 88f.) when discussing his relationships with the leading soldiers, Liddell Hart's impact on the problem of the Army's role was particularly strong in 1937 and 1938. His strategic outlook was in harmony with Neville Chamberlain's, and was widely shared in Government circles. When Chamberlain became Prime Minister in May 1937 he was already bent on settling the order of the Army's priorities which had been discussed indecisively throughout the previous year. Neither Inskip nor Hore-Belisha had the slightest grasp of strategic matters, and the Prime Minister needed an expert to back his own views against the obstinate insistence of the Chiefs of Staff that the Army must be prepared for a European war. From May 1937 to February 1938, when the Continental commitment was virtually abandoned, Liddell Hart supplied the draft policy papers which Hore-Belisha placed before his military advisers. Liddell Hart then criticized the latter's memoranda and caused—via Hore-Belisha—the final version to be rewritten several times before it went before the Cabinet. In the summer of 1937 he also advised Sir Thomas Inskip on the question of the role of the Army.[36]

Chamberlain endorsed the limited liability concept in his only letter to Liddell Hart (in March 1937) in which he wrote:

> I am sure we shall never again send to the Continent an Army on the scale of that which we put in the Great War.

Equally revealing is the Prime Minister's suggestion in a letter to Hore-Belisha of 29 October 1937:

> Among other things, I have been reading *Europe in Arms* by Liddell Hart. If you have not already done so, you might find it interesting to glance at this, especially the chapter on the role of the British Army.[37]

Thus for about a year from May 1937 Liddell Hart was truly the *eminence grise* behind the War Minister, but it was a precarious and unenviable role whose drawbacks quickly became apparent. Not surprisingly Gort, Adam and others whom Liddell Hart regarded as friends cooled rapidly in their attitude towards him when they realized that he was unofficially briefing Hore-Belisha 'behind their backs'.[38] In any case Liddell Hart was far too independent-minded and individualistic to remain long in a position which inevitably, if unfairly, associated his own name with Government failures. By the spring of 1939 he had become thoroughly disillusioned with Chamberlain and his Government for doubling the Territorial Army, introducing conscription and giving an impractical guarantee to Poland. The next chapter will illustrate his critical attitude to British policy and grand strategy during the Second World War. Though important politicians continued to consult him, he was never again to play such a direct part in the making of military policy.

In evaluating the strengths and weaknesses of Liddell Hart's military thought in the second half of the 1930s, it is essential to keep in mind the historical context. Like so many of his generation, Liddell Hart could not escape from the terrible effects of the First World War, partly from first-hand experience but perhaps even more from subsequent study and reflection on the errors and futility of its conduct. It is understandable that he and many of his contemporaries should wish to avoid a repetition of that holocaust at almost any cost—particularly if, like Liddell Hart, they believed that Britain could have avoided a full military commitment in 1914 comparatively easily and safely. Secondly, despite his confident forays into foreign policy and grand strategy, Liddell Hart's real expertise lay in the technical, tactical and operational aspects of war. This was the sphere in which he had campaigned so strenuously ever since the First World War and in which the backwardness of the British Army was so painfully obvious. In the later 1930s even Liddell Hart's skill with the pen could not conceal the conflict within his personality between the ardent tactician delighting in new operational techniques, and the humane, hypersensitive philosopher who loathed the violence, destructiveness and ultimate irrationality of war. Thirdly, in understanding his position it needs to be stressed that—though inclined towards isolationism—he was emphatically not an advocate of appeasement, but found himself writing for the most politically influential newspaper, whose editors certainly were. His behaviour in this dilemma was entirely honourable but his reputation undoubtedly suffered from the ambiguous relationship. Finally the basic fact needs to be underlined that he was a journalist who had to publish regularly to earn his living. This is not to suggest that had he been able to wait upon events his main ideas would have been significantly different; it is likely however that they would have been more carefully and coherently presented than in the articles and books which were hastily thrown together during his hectic years with *The Times*.

Given these extenuating circumstances, what are the main criticisms to be made against the leading proponent of limited liability? Liddell Hart was a very clever man and a truly gifted journalist, but he was not as dispassionate and 'scientific' as he liked to imagine. Previous chapters have revealed his preoccupations and prejudices, which were indeed obvious to some of his contemporaries. His remarkable self-confidence and air of authority in military matters, though in some ways an asset, did lend an air of contempt to his comments on military leaders, but he was genuinely surprised when this was pointed out.[39] As a corollary of this point, it could be said that he made far too little allowance for strict War Office accounting and, even more important, production problems in criticizing the conservatism of the British military hierarchy. Liddell Hart—and David Low—were correct in believing that 'Colonel Blimps' did exist in high office, but the diaries and biographies of those working inside the defence institutions, such as Hankey, Pownall and Ismay, provide an essential corrective to the heroic saga of 'progressives' versus 'reactionaries' as described in Liddell Hart's *Memoirs*.

Another defect which this chapter has tried to expose is that Liddell Hart's approach to the formulation of national defence policy was weighted excessively towards tactical considerations. Thus for example, he discussed the crucial issue of Britain's possible contribution to Continental allies almost entirely in terms of her inability to participate in mobile operations with mechanized units. He, and those who agreed with him in the Government, tended to understate the political and moral value of a definite undertaking to share in the ground defences of Belgium and France.[40]

It is far from easy to answer the question whether Liddell Hart accurately predicted the trend of military events in 1939 and 1940. It is certainly true that his writings contain numerous general references to the importance of an indirect approach, of psychological dislocation and deep penetration of the enemy's defences to paralyse his 'nerve system' of command; more specifically there are references which suggest that he saw the Ardennes as a possible unexpected approach route to France. Nevertheless a close study of his publications between 1935 and 1939 leaves a contrary impression that he was much more concerned that Britain and France might lose a war through launching a reckless offensive than that Germany could gain a decisive victory through attacking first. As Luvaas observes: 'Gort's prediction of the possible German breakthrough in France came closer to the facts than LH's at this time' [1937].[41]

A more curious limitation is that he does not seem to have grasped that economic pressures, as well as the ideological frenzy prevalent in Nazi Germany, would inevitably manifest themselves in the style of warfare adopted. Instead he argues that so superior is the defensive that it will be almost impossible for the aggressor to achieve decisive results. At best he hoped for an eighteenth-century type of limited war but employing mechanization to reduce casualties. Did he perhaps, as Michael Howard has suggested, know that Britain's allies were likely to be defeated and write 'partly to reassure himself, partly to reassure others' and above all to warn against a return to the suicidal strategy of the First World War?[42] More likely he was simply mistaken about the nature of the impending war. Whatever his private feelings may have been, in public he predicted the superiority of the defensive and the feasibility of limited war. It was thus inevitable that his reputation should suffer a decline after the events of 1939 and 1940, though some contemporary critics were grossly unjust in virtually blaming him for Dunkirk.[43]

The final reflection is that by denying the need for a Continental commitment on land, Liddell Hart was in practice undercutting his own Herculean efforts to reform the Army. By keeping the European role in doubt during the years 1934–37 inclusive and then relegating it to the lowest priority in 1938, the limited liability school in effect gave precedence to the competing claims of the other services and in particular to the air defence of Great Britain. How much difference, if any, an early decision to prepare the Army for European war would have made to the events of 1939–40 cannot, of course, be known with certainty.

It is difficult to resist the conclusion that had Liddell Hart been truly disinterested in outlook he could have retained all his views about the weakness of the Army and yet have accepted that a Continental commitment was unavoidable. Michael Howard's verdict on this unhappy period is severe but just:

> Nobody stressed more often the need for ruthlessly dispassionate analysis as a basis for both history and theory; but he himself sought to escape from the dilemma of his generation by what was, in the context of his times, little more than rationalisation of nostalgic wishful thinking. [To that extent his] eclipse was not undeserved.[44]

Notes

1. See Liddell Hart *Memoirs* II *passim*; J. Luvaas *The Education of an Army* (U. of Chicago 1964: Cassell 1965) pp. 406–407.
2. *Memoirs* II p. 118.
3. See the chapter 'Maginot and Liddell Hart: the Doctrine of Defense' by Irving M. Gibson (pseudonym for Arpad V. Kovacs) in E. M. Earle (ed.) *Makers of Modern Strategy* (Princeton U.P. 1941: O.U.P. 1944). It is regrettable that this misleading account should have been republished unrevised in subsequent editions.
4. This account assumes a general knowledge of the development of the 'Role of the Army' issue. For good concise summaries see Michael Howard *The Continental Commitment* (Temple Smith 1972) pp. 98–117; Brian Bond (ed.) *Chief of Staff: the Diaries of Lieutenant-General Sir Henry Pownall* I (Leo Cooper 1972: Shoe String Press 1973) pp. 126–129 (all references are to this volume).
5. In *The Times* see, for example, 'The Army of Today' (25, 26, 27 November 1935); 'Air, Land and Sea' (leading article on 10 February 1936); 'The Army Under Change' (30 October and 2 November 1936); 'The Army Estimates and the Army's Role' (5 March 1937); 'Defence or Attack?' (25, 26, 27 October 1937); 'The Field Force Question' (17 June 1938); 'An Army across the Channel?' (7 and 8 February 1939). See also *When Britain Goes to War* (1935) and *Europe in Arms* (1937).
6. *Memoirs* II p. 146.
7. R. J. Minney *The Private Papers of Hore-Belisha* (Collins 1960) p. 54.
8. *Europe in Arms* pp. 130–140.

9. *Memoirs* I pp. 293–294. *Chief of Staff* op. cit. p. 172. Burnett-Stuart Papers and letter published in *The Times* 28 November 1938. Colonel R. P. Pakenham-Walsh noted in his diary on 26 January 1937 'There seems to be a general opinion that Neville Chamberlain, basing his views on some loose talk by Jock Burnett-Stuart and articles in *The Times* by Liddell Hart, sees no use for a Field Force and hence the difficulty of getting anything for the army, as they are waiting now till he becomes Prime Minister when Baldwin goes' (Pakenham-Walsh Diaries). Deverell and Haining (then D.M.O.) also tried to convince Liddell Hart that his newspaper articles would have the effect of impeding Army reform. This 'Talk with F.M. Sir C. J. Deverell' 12 November 1936 (11/1936/99) epitomizes the differences between General Staff thinking and Liddell Hart's.

10. Most of these points were made in *The Times* article 5 March 1937. See also Liddell Hart's letter in *The Times* 11 November 1936 on 'The Role of the Army' and General Pritchard's reply.

11. Pakenham-Walsh Diary op. cit. 26 March 1937.

12. See for example Cab 53/27/460 (J.P.) an Appreciation by the Joint Planning Sub-Committee of the C.I.D. on 29 April 1936 of an outline plan to dispatch two divisions to the Continent 14 days after mobilization. They considered placing the Force (a) between the French and Belgians (b) on the Belgians' left or (c) in reserve behind the left flank in Belgium—which they preferred. It should be noted that in all these discussions, *pace* Liddell Hart, there was no suggestion of the B.E.F.'s participating in a general offensive.

13. See for example, the Chiefs of Staff's discussion on 4 May 1934. Cab 53/4/125.

14. *Memoirs* I p. 298.

15. 'Defence or Attack?' *passim. Europe in Arms* pp. 296–297; repeated in *The Defence of Britain* p. 50.

16. Talk with Deverell 18 November 1937, 11/1937/94. Reading between the lines, it is easy to see that the C.I.G.S. became confused and heated under Liddell Hart's persistent questioning. Pressed on where he imagined a complete Field Force could operate offensively he 'said vaguely, "places like Libya" '.

17. For useful references to journal articles on the defence-or-attack question and related issues see Gibson's chapter in *Makers of Modern Strategy* op. cit. pp. 380–382.

18. *Chief of Staff* op. cit. pp. 99, 139–140.

19. *Army Quarterly* XXXVI (1938) pp. 123–127, cited in *Makers of Modern Strategy* p. 381.

20. *Sunday Express* 9 July 1939. Another article in the *Sunday Dispatch* on 21 May 1939 entitled (against Liddell Hart's wishes) 'Why Hitler and Mussolini Could Not Win a War' at least proved correct in the long run.

21. *Memoirs* I pp. 285–288.

22. 'Note on the Reoccupation of the Rhineland Demilitarized Zone' 7 March 1936, 11/1936/45. 'Talk with Paget and Adam—Athenaeum' 15 May 1936, 11/1936/64.

23. *The Defence of Britain* pp. 63–64. *Memoirs* II pp. 127–130, 188. Eden agreed with Liddell Hart's view that if Franco won he would probably be

pro-German and would threaten Britain's trade routes (holograph note) 1 June 1938, 11/1938/62.

24. *Memoirs* II pp. 160–172.

25. Ibid. pp. 193–194. *Chief of Staff* op. cit. p. 151. For an early admission by Liddell Hart that a Continental commitment was necessary see his Diary note on 29 January 1939 (11/1939/6).

26. *Memoirs* II pp. 197–198. *The Defence of Britain* p. 209. Liddell Hart's Diary notes (typed extracts) 30 January 1939. 11/1939/6. Spears to Liddell Hart 7 February 1939. In his reply on 17 February Liddell Hart wrote 'I find that there are a number of people in the Government and connected with it, who were brought gradually to accept the case for a return to our traditional war policy, and are not very willing to admit that by the Munich settlement they undermined its practicability'.

27. Copy of Burnett-Stuart's letter to *The Times* in his Papers, undated but published 4 March 1939. R. Macleod and D. Kelly (eds) *The Ironside Diaries, 1937–41* (Constable 1962: *Time Unguarded* McKay 1963) p. 48.

28. *Chief of Staff* op. cit. pp. 123, 129.

29. Liddell Hart holograph note 3 June 1937, 11/1937/49. Talk with Sir Cyril Deverell 29 June 1937, 11/1937/56. Talk with Brigadier G. Le Q. Martel 28 November 1938, 11/1938/122. Pakenham-Walsh Diary 16 December 1936.

30. Cab 21/510 Hankey to Inskip 7 March 1938.

31. 11/1936/116 (J. de la Ferté), 11/1937/34 (Sir Donald Banks). On 30 September 1938 Major-General Sir F. A. 'Tim' Pile told Liddell Hart that according to Air Staff calculations the Germans could drop a maximum of 600 tons of bombs per day on Britain; 60,000 would be killed in the first week but pressure would decline after a month: 11/1938/105. Another close R.A.F. contact was Air Vice-Marshal R. H. Peck—see for example 11/1939/56.

32. Liddell Hart to Ironside 11 March 1937.

33. 11/1937/38 and 99.

34. 'Outline of the Opposition to the Development of the Anti-Aircraft Defence of Great Britain' 30 July 1938, 11/1938/89. 'Offensive versus Defensive in the Air' 26 May 1938, 11/1938/60.

35. Liddell Hart was in close touch with prominent opponents of appeasement, including Churchill, through the 'Focus' group: see 11/1939/10. For a talk with Lord Halifax see 11/1936/29, and for an important memorandum sent to Halifax on 8 September 1939, 11/1939/100. (Also F.O. 800/317, 43A.) Lord Lothian particularly sought Liddell Hart's advice on strategic matters before becoming British Ambassador to the United States; see his letter dated 24 July 1939 (*Memoirs* II p. 248).

36. *Memoirs* II pp. 50–104 *passim*. For Liddell Hart's advice to Inskip see 11/1933/56. How Liddell Hart's repeated insistence on the re-drafting of the 'Role of the Army' paper was received by the General Staff can be seen in *Chief of Staff* op. cit. pp. 129–133.

37. *Memoirs* I p. 386, II p. 39.

38. *Memoirs* II pp. 19, 87–88, 250. *Chief of Staff* pp. 138, 148, 190. J. R. Colville *Man of Valour: Field Marshal Lord Gort V.C.* (Collins 1972) pp. 82–87, 95–98. Hobart later told Liddell Hart that when he was discussing training

ideas with Gort and Adam in February 1938 the former had remarked 'We do not want any more of Liddell Hart's ideas—there have been too many of them already': 11/1940/5.

39. As Burnett-Stuart wrote to Liddell Hart on 20 September 1932 (concerning *The British Way in Warfare*): 'Imagine yourself the responsible head of the Army and read your book from that standpoint. Wouldn't you rather resent your own accusations of incompetence and rather contemptuous assumption of superiority? And wouldn't all sorts of difficulties confront you, which as a critic you make no allowance for?' See also Luvaas op. cit. pp. 420–421.

40. With a few honourable exceptions such as Duff Cooper, Eden and—eventually—Halifax, Cabinet ministers did not seem to realize that Britain's fate was ultimately linked to that of France; indeed in discussing 'the Role of the Army' they almost seemed to invite the charge of 'perfidious Albion'. See, for example, the C.I.D. discussion on 15 December 1938 (Cab 2/8).

41. 'May it not be possible for panzer divisions and concentrated air forces to effect a breach and this attack can take place with little previous warning. If by rapidity, deception and surprise it is possible to make a bridgehead then the war will pass into open country once more. I feel novelty lies in some such direction as this as Belgium is hackneyed.' Typical Liddell Hart? In fact these are Gort's words and Liddell Hart thought his suggestion improbable because mechanization would strengthen the defensive. (Correspondence on the 'Defence or Attack?' articles in Gort file 1937. Luvaas' comment is taken from his typed notes on this correspondence.) In *The Defence of Britain* pp. 217–219 Liddell Hart also saw the Ardennes as an asset to the defenders rather than as a covered approach for the attacker.

42. Michael Howard 'Liddell Hart' in *Encounter* April 1970.

43. See the chapter in *Makers of Modern Strategy* cited above. Liddell Hart's unpopularity at the beginning of the war will be referred to in the next chapter.

44. Michael Howard in *Encounter* op. cit. p. 42.

5

The Critic of Total War

In professional terms the Second World War was a period of disappointment and frustration for Liddell Hart. On 10 September 1939 he wrote: 'As I am unable to do anything effective now to check this fatal course, I do not care to be an accomplice in the vain sacrifice of the nation's youth, and the wrecking of British civilisation.' He relinquished his influential post of defence correspondent for *The Times* in the autumn of 1939, partly because he had disagreed with the paper's editorial policy, but also because he took a strict view of the need to tell the truth in wartime and did not believe that he would be allowed to do so in *The Times*. This scruple severely restricted his outlet as a freelance journalist in the early months of the war, but from March 1941 onwards he was the regular military correspondent of the *Daily Mail*. He was offered no official appointment during the war and remained very much an outsider.

In personal terms his health was slow to recover from the heart attack, brought on by overwork, which he had suffered in June 1939. Furthermore his sudden drop in income posed an acute dilemma between his desire to preserve his integrity as a writer and his need to earn a living. His friends Dorothy and Leonard Elmhirst came to the rescue by offering him the use of part of a house on their Dartington estate at a modest rent, so that his operational base remained in Devon until June 1941, when he moved to High Wray House near Ambleside in Westmorland. This proved to be a happy haven for the rest of the war. 'After being without a place of my own since 1938,' he wrote to a friend, Vivian Gaster, 'I have taken a most delightful house (with vegetable garden, orchard and even wild strawberries "on the estate").' However it should not be imagined that he led a sedentary, bucolic life in these two idyllic settings: on the contrary his diaries reveal a life of constant travel, mainly in the form of frequent forays to London and regular tours of the military training areas. Indeed one of the advantages of his new home was that it was 'conveniently close to the training ground of our new armoured forces in the North, commanded by a friend of mine [Major-General P. R. C. Hobart] who is a refreshing contrast to the ordinary run of generals'. His domestic life entered a happier phase when, in 1942, his divorce proceedings being at last completed, he married Kathleen Nelson, who henceforth provided an ideal companion and helpmate in his work.

At the end of the last chapter it was remarked that with the outbreak of the Second World War Liddell Hart's public reputation suffered an eclipse. Some of the reasons for this will be made clear in the present chapter and here we need only mention that as 'Appeasement' became a dirty word his name was (unjustly) linked with *The Times*, which had supported the policy; and also that his emphasis on the superiority of the defensive and the relative impotence of the offensive was seized on by many critics as a monumental misjudgement when Germany overran the West in May and June 1940.[1]

Liddell Hart also suffered an eclipse in another sense in that many of his avenues of influence to service chiefs and politicians in power were either completely closed or at least temporarily blocked. Ironically, some of the generals whose promotions he had helped to secure, including Gort and Adam, were now opposed to his being consulted. In short, by contrast with the late 1930s, during the war years he was operating only on the fringe of politics. Not only did he hold no office or regular advisory post, but the only major politicians with whom he was in frequent contact were Lloyd George (a rapidly waning star after 1940) and Beaverbrook, who finally left the War Cabinet in February 1942. Had either succeeded Churchill as Prime Minister, as briefly seemed conceivable, then Liddell Hart would probably have been invited to become an official military adviser.

In the development of Liddell Hart's thought the war years are paradoxically among the most interesting yet little-known of his whole career. The second volume of his *Memoirs* has only a short concluding section taking his story from the outbreak of war to the fall of France. Jay Luvaas' well-informed essay also stops in 1940 with merely a glance into the future; while the somewhat misleading chapter in E. M. Earle's *Makers of Modern*

Strategy was written early in the war and never revised. More recently, interest has focused either on Liddell Hart's influence on the development of armoured warfare or on aspects of his career after 1945. In the present writer's experience he seldom spoke spontaneously about the war years.

This is not to suggest that he felt he had anything discreditable to conceal, and indeed he had initially planned to take his *Memoirs* up to 1945, until they got impossibly lengthy. On the contrary, his consistent criticism of the waging of total war in the vain pursuit of the mirage of victory, though it made him something of a heretic in official circles and effectively debarred him from playing an active role in the war, required immense integrity and moral courage. Moreover, in the longer perspective of post-1945 disillusionment and with the onset of the cold war, Liddell Hart's wartime urging of restraint towards the enemy and his pessimism as to the possibility of a meaningful 'victory' appeared in a more favourable light.

The course of his thinking during the war is extremely well documented in his files, so much so that the problem is what to select among numerous interesting letters and memoranda. Here again Liddell Hart's scrupulous regard for preserving the complete historical record, including some telling criticisms of his basic ideas about the war, must be held in his favour. Thus the historian can readily discover not only what Liddell Hart himself thought at any time, but also to whom he circulated his various reflections and memoranda, and what replies he received. In short, he has preserved an invaluable source for the critical, anti-establishment, anti-Churchillian view of the Second World War, an aspect which will surely in time receive much more attention from historians.[2]

Liddell Hart's standpoint in the early months of the war appeared unorthodox yet was widely held in Parliament and the country. After the outbreak of war he proposed conciliation and a negotiated settlement just when the British Government—at any rate on the surface—and much more the Press, were renouncing the policy as inapplicable to Nazi Germany. Having advocated collective security in the mid-1930s and criticized the Munich settlement, Liddell Hart nevertheless deplored the British Government's hasty reactions to the Nazi seizure of Czechoslovakia in March 1939. He was a bitter critic of the adoption of conscription on both military and moral grounds, and he regarded the guarantee to Poland without an accompanying understanding with the Soviet Union as the utmost folly and an open provocation to Hitler. Thus he believed that Britain and France had chosen an emotional and unconvincing cause for which to go to war. Furthermore, he was sceptical on military grounds of the prospects of offensive operations: in retrospect he was to emphasize that he had been correct in the case of Britain and France, but an impartial reader of his writings before May 1940 will surely deduce that he was discussing the offensive in general and had underrated German, not to mention Japanese capabilities.

For all these reasons Liddell Hart was appalled at the irrational surge of jingoist optimism in support of an all-out war effort to secure an early victory over Germany which swept the country after the declaration of war in sharp reaction to the rather shame-faced acceptance of appeasement. Some of his own rather surprising ideas, favouring a negotiated peace for example, must be viewed in the light of his large collection of wild, aggressive and even hysterical outbursts. As early as 10 September he noted that Lloyd George and many of the papers were proclaiming 'no negotiations with Germany'. On 2 October 1939 Churchill, then First Lord of the Admiralty, provoked Liddell Hart's criticism by promising eventual victory in a radio talk. On 13 November Churchill made a 'fighting' broadcast declaring that either we would be beaten or Nazism and the German menace to Europe would be *broken and destroyed*. 'Churchill,' Liddell Hart commented, 'uttered a string of phrases which made Hitler's speeches seem almost polite.' Unfortunately for Liddell Hart, Churchill's rhetoric about Nazi barbarism was to be proved accurate. He noted a list of silly statements by Churchill, Ironside (then C.I.G.S.) and Chamberlain, including the latter's assurance on 4 April 1940 that 'Hitler has missed the bus'. On 18 February 1940 Hore-Belisha, who had recently been dismissed from the War Office, advocated that Britain attack Russia in Finland and go on to seize Leningrad. Admiral Sir Roger Keyes and F. A. Voigt also urged that Britain should attack both Germany and the Soviet Union. John Gordon's articles in the *Sunday Express* provided many choice examples of the extremism which Liddell Hart sought to curb. Thus on 18 August 1940 the former wrote 'They who have boasted that they were without mercy deserve no mercy . . . For every bomb dropped on the soil of Britain . . . drop 10 bombs into the heart of Germany.' Others named in Liddell Hart's 'Roll of Dishonour' included Reynaud, Churchill, H. G. Wells and Lloyd George (for their deplorable reactions to Belgium's capitulation), Bertrand Russell ('for abandoning pacifism and applauding the war from New York') and 'Cassandra'—William Connor of the *Daily Mirror*, who later became a friend—('for his disgusting broadcast on [P. G.] Wodehouse'). This, it should be stressed, is only a small sample from the files.[3]

From the outset Liddell Hart consistently opposed the vigorous waging of war and urged instead that Britain should press for a negotiated peace. In a widely-circulated paper dated 8 September 1939 entitled 'The Need for a New Technique of War', he argued that while Britain and France could not defeat Germany they had certain advantages over her, among them that the German people were ill-fed and under strain.

> To pursue a military offensive against Germany would be the most unwise strategy and policy from every point of view . . .
>
> A declaration that we are renouncing military attack as a means of combating aggression would be a far sighted move, strengthening our moral position, while forestalling the otherwise inevitable growth of derision abroad and disillusionment here. It might well be the first point of a new technique of war, suited to present conditions and our particular circumstances. It would set us free to develop economic and moral pressure to the utmost and to make the best disposition of our military forces to meet any German attempt to break our 'sanitary cordon'. It would throw on the Germans the responsibility of taking the offensive, with all its disadvantages.

He also thought that such a non-bellicose policy might be better calculated to secure the co-operation of the United States.

A few days later he elaborated the argument in a memorandum entitled 'A Personal Problem':

> To seek victory in a modern 'great war' is never more than the pursuit of a mirage in the desert. In the present war it may be worse—leading us into a bottomless quicksand not merely in the national sense. For the prolonged struggle made inevitable by such a

fallacious aim endangers the existence of all spiritual values. Freedom itself may hardly survive such a reckless offensive 'defence'.

He looked for signs that statesmanship had freed itself from the illusions of 'victory'. He advocated doing nothing to antagonize Germany in the hope that she would fall out with Russia. 'This war,' he wrote, 'may be decided by boredom rather than blood.' He even deplored the 'heroic' defence of Warsaw because it added to the destruction to no good purpose.[4] This was a very detached observation indeed, since it ignored the value of myth-making and hope for the future even in apparently total defeat. Not surprisingly, he had difficulty in publishing his views on the folly of active war at the very moment when the Government had at last decided to take the plunge. *The Times* declined to publish 'The Need for a New Technique in War', but he included this and other rejected articles in his book *The Current of War*, which appeared on 27 February 1941.[5]

In order to believe that a negotiated peace with Hitler was possible after his conquest of Poland, Liddell Hart had to take the view that the Führer was essentially cautious and limited in his territorial ambitions. Long before Mr A. J. P. Taylor's challenging revisionism, Liddell Hart developed the conviction (obsession would not be too strong a word) that Britain had in effect 'caused' the outbreak of war by giving a provocative guarantee which she was powerless to implement. Germany had a strong case over Danzig and the Corridor and we should have negotiated on it as Mussolini suggested. We should have pressed the Poles to accept an accommodation with Hitler before the crisis came to a head. Moreover Hitler had good reason to feel provoked by the Poles and by the foolish Anglo-French guarantee. Thus 'an irresistible temptation was combined with an indefensible provocation—indefensible, that is, on the part of statesmen dealing with hard facts'. By 1941 Liddell Hart could write 'Even more than most wars, this war has, from the outset, made no sense'. 'Easy to blame it on Hitler—if you push your hand through the bars of a tiger's cage, and get mauled, anyone realises it is your own stupidity. Especially if you hold out a steak, pull it away and hit him on the nose.' In *Why Don't We Learn From History?* (1944) he extended the notion of 'the inherently war-provoking effect of the

Polish Guarantee' to the wider charge that the natural result of our 'offensive' talk had been to *precipitate* German intervention in Scandinavia, the Low Countries and the Balkans. Thus responsibility for the consequent misery that had befallen the peoples of Denmark, Norway, Holland, Belgium, France, Yugoslavia and Greece in turn 'lies heavily upon us—for losing the sense of military realities'.[6]

What is one to make of all this? Criticism of the Polish guarantee was certainly justified on the grounds that it was hastily undertaken without proper military consultation, it was virtually impossible to implement militarily, and it was not promptly followed by a really serious attempt to gain a Russian alliance despite the unequivocal advice of the Chiefs of Staff. Furthermore, it probably did to some extent 'provoke' Hitler, though he was already known to have designs on Danzig and the Corridor. On the other hand, it cannot be said that Liddell Hart displayed any imaginative understanding of the British Government's difficulties or of the motives behind the guarantee. As Maurice Cowling has written:

> Chamberlain did not intend the Polish guarantee to exclude a frontier revision. On the contrary, he wanted the Poles to be reasonable. While hoping that they would be, he felt more confident than the Foreign Office that Hitler knew that Britain would go to war if they were attacked. He thought even Ribbentrop understood this, and that Hitler, who 'was not such a fool as hysterical people make out', would be willing to compromise 'if he could do so without what he would feel to be humiliation' . . . It was in order both to be flexible and to make Hitler flexible that he [Chamberlain] permitted a renewal of the economic discussions which Prague had halted.[7]

In other words Chamberlain was attempting to do what Liddell Hart later advocated, namely 'to run risks of war for the sake of preserving peace'. Whether or not Hitler felt he had been provoked, the fact is that he opted to get his way by war, as he had wished to do at Munich, without waiting to see if Britain and France would once again hand him what he wanted 'on a plate'. This example of the Polish guarantee aptly illustrates a strange quirk in Liddell Hart's mentality which caused him to be severely critical of British (and to some extent Allied) policy and actions while making every allowance for the enemy. The same trait will be discussed later in this chapter in relation to other aspects of the war such as strategic bombing and atrocity rumours.

As a corollary to his opposition to the vigorous prosecution of the war for military reasons, Liddell Hart also feared that the more 'total' the war became, the greater the risk that 'freedom' would be permanently lost. Although he wrote eloquently and often about 'freedom' he seldom examined the concept in any detail. Clearly what he chiefly had in mind was personal liberty from bureaucracy and compulsory service, and freedom to publish critical or unorthodox opinions. What is surprisingly absent in his writings at this time is the more obvious keynote, sounded by Churchill and others, that unless Nazism could be defeated a great part of Europe would remain permanently subjected to a barbarous tyranny where personal freedom would entirely disappear. As early as October 1939 he feared that Britain was already becoming a totalitarian state—'our blind leaders are blindfolding the people'. He was appalled at the way in which alien refugees from Nazism, among them artists and scientists of distinction, were rounded up and interned. It was a sign that 'the Fascist gang are taking over here'. He also noted cases of wild talk from officers on leave of shooting 'conchies', 'huns' etc.[8]

Not surprisingly, as a working journalist, he was particularly sensitive to the restricted freedom to publish unpopular opinions in the Press, though it must be said that this was more due to editors' reluctance to encourage 'alarm and despondency' than any official censorship. In an awful pun he reflected on 8 October 1939 that 'the Press ought to be named the Suppress'. Lloyd George, who suffered from the same treatment from editors as Liddell Hart, and for the same reason, plus his sweeping criticism of Churchill, commiserated on 20 November 1939:

> Express have [*sic*] done me the honour of putting me in the same category as yourself as a man who tells too much of the truth about the war to be encouraged in their columns. I am afraid that the realists are in for a bad time for at least some months but I think our chances will come when the nation begins to understand the utter futility of this ghastly struggle.

In September 1941 Liddell Hart told Lloyd George that the *Daily Mail* had cut an article in which he urged that the German people must be offered a way out of the present situation. The former published several of these unpalatable articles in his wartime books with accompanying notes of when and where they had been rejected. He found the French official censorship even more restrictive.[9] These, however, were minor irritations compared with what he would have encountered in Germany or the Soviet Union.

The threat to freedom which Liddell Hart regarded with the greatest horror was conscription. In a holograph note entitled 'A Reflection on Freedom' dated 7 August 1941 he wrote of conscription:

> The effects far transcend the military sphere. Bemused by the cry of total warfare, we are trying to make ourselves totalitarian—with the maximum of inefficiency for the minimum of productivity, in proportion to the effort ... The basic principle of Nazism is the claim of the State to determine the individual's duty, and decide his conscience for him. Hence, in opposing the Nazi's 'New Order', we weaken our own position if we adopt the same basis....

In an expanded note he added that we should be careful to protect the rights of conscientious objectors and other non-conformists.

In the following year he developed a variation on the conscription theme based on the premiss that 'civilised peoples are not good fighters—we ought to recognise this and adapt the lesson to our needs'. Conscription in a civilized state is apt to weaken the army by drawing in more and more non-aggressive people. Those who frame the defence policies of civilized nations would be wise to constitute their forces of good 'fighting animals' as Britain has done by drawing on the 'martial races' in India. Here was another way back to the limited warfare of the eighteenth century.

Napoleon's responsibility for instituting conscription was referred to in a reflection dated 30 January 1943:

> Napoleon fell, but left as a legacy the chain of military conscription—which dragged mankind into a series of bigger and badder wars. When Hitler falls, will he also leave the chain of civil

> conscription, the logical corollary of totalitarianism rivetted round the necks of mankind—thus establishing the reign of universal servitude? If so, it will be an ironical reflection on the unthinking conduct of the war, and on the efforts and sacrifices made by the peoples who have sought and fought to defeat him.

In a long, pessimistic section in *Why Don't We Learn From History?* Liddell Hart went even further and, overlooking the possibility that conscription might be the *effect* of broad social, economic and political developments, characterized it as 'the greatest contributory factor to the Great Wars which had racked the world in recent generations ... Conscription serves to precipitate war, but not to accelerate it—except in the negative sense of accelerating the growth of war-weariness and other underlying causes of defeat.' As regards Britain in the Second World War:

> A grave responsibility is borne by the members of a Parliament which allowed, and even encouraged, officialdom to impose such unconstitutional measures while at the same time turning this country into a potential prison-house, from which in case of defeat or a *coup d'état* it might be impossible for anyone to escape and start a 'Free British' movement.[10]

What is striking about these comments on conscription, which reappear throughout his writings, is the dogmatic nature of their formulation. Undoubtedly there was much truth in Liddell Hart's views: for example, the existence of conscript armies was indeed a factor that helped precipitate war in 1914. Yet once an idea like this has taken root it tends to be developed and endlessly repeated without considering possible arguments on the other side. In the Second World War conscription probably was a crude and uneconomic way of manning armies, but governments had to contend with both the *moral* problem of equality of sacrifice and the *manpower* problem of attempting to distribute the skills of the nation for the most efficient running of the war machine as a whole. However ineptly this may have been done it does not obviate the basic problem. In this and other matters Liddell Hart sometimes seems to be a skilful debater rather than the objective seeker of truth on which he prided himself.

By the late autumn of 1939 Liddell Hart was even more

pessimistic than he had been at the start of the war about Britain's prospects of victory, and his files contain numerous references to disputes with optimists. On 1 November he noted 'The most senseless of our wars in the eighteenth century was christened "the War of Jenkins's Ear". The present war may come to be known as "The War of Chamberlain's Face".' A week later he summed up his views in a memorandum entitled 'The Prospect in this War', copies of which were sent to Lloyd George, Eden, Dalton, Cripps, Toynbee, Lord Cecil, Attlee, Hore-Belisha, Sir Arthur Salter and others. He outlined the ways in which Britain could lose the war and stressed the difficulties in winning, but he concluded that the most likely outcome was a stalemate in which all would lose. If this was to be avoided one party, as in a domestic quarrel, had to give the other a real chance to climb down without loss of face: 'A perception of these underlying realities should make it easier for us to proffer a ladder by which Hitler can climb down if he shows any disposition to do so.' It should be emphasized that, like Lloyd George, Liddell Hart was never a 'defeatist' in that he never for a moment considered Britain's independence to be negotiable; rather, he assumed that *Hitler* would wish to open negotiations once he realized that Britain was impregnable. Most of the replies were non-committal but Lord Cecil agreed that we should make every effort to achieve a negotiated peace consistent with our honour. The difficulty was that he could not see how we could do this short of requiring Germany to evacuate Poland and Czechoslovakia as a preliminary to discussing what should be done with those territories. This was exceedingly unlikely to happen, but even Lord Cecil did not mention the problem of obtaining the withdrawal of the Soviet Union from the portion of Poland which she had occupied.[11]

In a similar memorandum entitled 'The True Object In War' Liddell Hart touched a keynote which was to be frequently repeated, namely that the dictators, though short-sighted and callous in pursuing their aims, at least wanted to avoid a ruinous war. By contrast the leaders of the democracies were now showing a reckless disregard of all consequences. We should face the unpalatable necessity of trying to get a compromise peace and show Germany it was in her interests as well as ours. Liddell Hart was not put off by the obvious snag that even if the

Germans could be encouraged to 'start bidding for peace', Hitler's word could not be trusted. Is our own Government any more trustworthy, he asked rhetorically, in view of its *volte face* over conscription? By publicly advancing such views Liddell Hart was of course courting criticism, and at least one writer, Major-General Sir Charles Gwynne, openly accused him of defeatism for denying that Britain could win. In March 1940 an official of the Ministry of Information wrote informing Liddell Hart that 'our critics in the United States are spreading the view that the Allies cannot win the war and that they are citing in support of this view alleged statements by you'. Would he be willing to help the war effort by writing something for publication in America to the effect that the Allies could achieve their war aims? Liddell Hart replied that, much as he would like to oblige, he could not see any reasonable grounds that the Allies could *win*; their best hope was to show the enemy he could not defeat them. He thought the best line to take regarding the United States was to stress *their* strategic dangers should the Allies be defeated.[12]

The early months of 1940 on the home front were dominated by Anglo-French proposals to distract Germany by opening up a new war theatre in Scandinavia. The Soviet Union's invasion of Finland provided a pretext and Germany's dependence on Swedish iron ore, shipped down the Norwegian coast in the winter months, offered an attractive strategic reason. Though it cannot be claimed that Liddell Hart foresaw the brilliant skills in combined operations which the Germans would display in Norway, he wisely counselled caution in opposition to enthusiasts for the venture such as Ironside and Hore-Belisha. For one thing he had no desire to see Britain in conflict with the Soviet Union, but more surprisingly he denied the value of an 'indirect approach'. He argued that we should avoid 'getting strategically frozen in trying to get a flanking grip on some remote part of the enemy's territory'. We ought to have learnt the lesson of such folly from the last war in diversionary operations like those at Salonika. This admission casts doubt on the general value of the theory of indirect approach. However he regretted that we had not given Germany back some of her colonies in 1918 so that now we could seize them again to show that aggression does not pay![13]

Throughout the weeks before the German attack on the Low Countries and France, Liddell Hart consistently opposed the prevailing ideas in the Press that we should go on to the offensive in Norway, bomb the Ruhr and generally go all out to win the war. He found prominent supporters for his views in Beaverbrook and Lloyd George. The former wrote frequently and fulsomely to Liddell Hart, more than once calling himself his 'disciple'. Early in March 1940, after expressing emphatic agreement with Liddell Hart's views, Beaverbrook urged him to go on a speaking tour to disillusion the public. Beaverbrook had been attacked for publishing Liddell Hart's 'defeatist views' in his newspapers but he thought they were making an impact.[14]

Liddell Hart believed that public opinion was increasingly turning to Lloyd George as the man to deal with the problem. When the Government encountered stiff Parliamentary criticism over its handling of the Norwegian operations, Liddell Hart urged Lloyd George to speak in the House of Commons, which he did with devastating effect on 8 May. In a diary note on that day Liddell Hart recorded that Lloyd George had expressed agreement with his views. The 'victory talk' by Greenwood and Sinclair in the debate of 7 May was nonsense: 'It was essential to be realistic, and while doing our utmost to foil the German aim, to work for an honourable peace.' He also recorded that Lloyd George fancied his chances of returning as Prime Minister: he had the best chance of overcoming Russian suspicions and he was the only man who could deal with Hitler and Mussolini on equal terms. Liddell Hart outlined a new war policy: we should build up our power of defence, make ripostes to the enemy's moves and introduce a social and economic policy designed to create a 'New Order' at home.[15]

Liddell Hart was disappointed to learn that Lloyd George was not included in Churchill's War Cabinet, though he had been offered a place. Frances Stevenson, Lloyd George's secretary, later informed Liddell Hart that the chief reason he had declined was because Churchill would not appoint a small War Cabinet whose members would be free from departmental responsibilities. On 11 May Liddell Hart noted gloomily that 'The new War Cabinet appear to be a group devoted to "victory" without regard to its practical possibility. Whatever the faults of Hitler and Mussolini, they have at least shown a limited understanding of the dangers of wrecking European civilisation ... But it is the democratic leaders who now threaten to set loose uncontrollable forces ...'

Liddell Hart continued to hope that Lloyd George would be recalled to office. In a talk on 23 July he told Liddell Hart that Sumner Welles, Roosevelt's Under-Secretary of State, had found the attitude in London the worst anywhere in regard to possible mediation or a peace settlement. Liddell Hart noted Lloyd George's pessimism:

> In the last war he had always been able to feel that we had the resources to win in the long run, so that he was justified in pressing on with the war without having to keep in mind the *necessity* of negotiating peace. But our policy in this war was a gamble and, he felt, a very dubious one. The best chance lay in trouble arising between Hitler and Stalin.

In an earlier talk (on 18 July) Lloyd George expressed the opinion that Hitler *would* attempt an invasion of Britain and he put our chances of success at only 5 to 4 on. Later in the summer, Tom Jones spoke privately to Liddell Hart at Criccieth about Lloyd George's pessimism and the way it was threatening to impair his influence. Lloyd George was very frightened of bombs, particularly when staying at his other residence at Churt in Surrey.[16]

Even before France's defeat was completed, Liddell Hart began to collect extracts from his inter-war writings to prove that he had accurately foretold what would happen, and to defend himself against the charge that in *The Defence of Britain* and elsewhere he had proclaimed the superiority of the defence over the attack.[17] There was of course much justification in this self-defence. He had usually been careful to qualify his statements, and it would have been difficult to predict that the French would commit so many errors as to nullify their strategic advantages:

> The French army paved the way for its own defeat because it failed to adopt or develop a defensive technique suited to modern conditions. It precipitated disaster, after the initial German penetration, by trying to apply an offensive technique that was utterly out of date. By contrast, the Germans succeeded because they had understood the strength of modern defence sufficiently well to avoid

> blunting their weapon in any frontal assault, while converting to their own advantage that same defensive asset in such a way as to make the Allies' counter-attacks recoil against themselves.

Nevertheless it is hard to resist the conclusion that Liddell Hart was protesting his accuracy too soon and too obsessively. In general his pre-war writings *had* left the impression that the attacker had little chance of success without an overwhelming preponderance in material strength, while since the outbreak of war he had emphasized the likelihood of stalemate which would only be endangered if the *Allies* took the offensive. In *Dynamic Defence* (published in November 1940) he first suggested that Hitler's triumphs in the spring were due not so much 'to the offensive power of his instruments' as to the initiative enjoyed by the aggressor, but a few pages later he referred to the Germans' 'vast superiority in tanks and aircraft'.[18] Later research has confirmed the Germans' superiority in the air, but their tank strength in both quantity and quality did not exceed or excel that of the Allies.[19] Liddell Hart, like Clausewitz before him, had strong grounds for believing the defensive to be *theoretically* stronger than the offensive, but the German *blitzkrieg* operations in Poland and the West demonstrated that *in practice* offensive operations *could* succeed in modern conditions—a lesson that was to be reinforced by Japan in quite a different environment in 1941–42. Liddell Hart would have done better to admit that his doctrine had been too sweeping, rather than try to explain away all these examples as exceptions. In *The Current of War* he indulged in casuistry to the point of arguing that 'In the usual military sense "the offensive" is a term applied, and restricted, to action which takes the form of advancing upon the enemy. This is a convenient, but not a precise description—especially in modern conditions of war ... A man, or a force, lying in a covered position may appear to have taken up a defensive attitude—yet his or its fire is operating offensively ...'[20] Quite so, but such abstract speculations were exceedingly remote from the practical problems confronting the military planners in the Second World War.

It might have been expected that the events of the summer of 1940, including Italy's entry and the consequent extension of war to the Mediterranean and Middle East, would have shattered Liddell Hart's hopes for a negotiated peace without victory. On the contrary he continued to hold consistently to his views, which now really were becoming unorthodox, and circulated them to as wide a group as before. His conviction of the futility of total war, derived from twenty-five years of study, was founded on six assumptions as follows:

1. Every war between great powers was avoidable.
2. There are always faults on both sides.
3. The common people are always the chief sufferers.
4. Suffering is increased by the pursuit of illusory 'victory'.
5. Victory is rarely attained in a military sense and never in a political sense—of being ultimately more profitable than an indecisive issue.
6. Even when victory is attainable it is a bad foundation for subsequent peace.[21]

In a paper entitled 'Simple Arithmetic For Statesmen' he argued that calculable factors showed the increasing possibility of an adverse result for Britain from continued war. Imponderable factors, though not to be discounted, required an extreme stretch of imagination to picture our *victory*.

> In securing a favourable peace settlement, however, we could count on the nuisance value of our air campaign and blockade—so long as we are capable of keeping them up. It would be fatal to delay such a settlement until a moment when our powers were declining.

In a similar memorandum dated 12 October Liddell Hart outlined his idea of a peace plan. Any settlement must leave Britain free to develop her air and sea strength—in the long run the growth of *American* air power would give us security and a stronger bargaining position:

> Then, when our combined strength was fully developed, we and the Americans might bring about any desired revision of the peace terms by the pressure of the most practically convincing of arguments—the possession of air superiority combined with control of the seas. To overcome Nazi domination of Europe we need a 'Ten Year Plan'—a long-term policy which, instead of pursuing the mirage of victory in a fight to exhaustion, will seek an opportunity of breaking away from the present clinch with a view to gaining the ultimate advantage in the most strength-conserving and surest way.

One of the few recipients of this memorandum to reply at length, voicing doubts about Liddell Hart's notion of a negotiated peace, was the novelist and editor of *John O'London's Weekly*, John Brophy. The latter admitted frankly that 'All my own instincts are against your conclusions . . . I loathe Hitler and Mussolini so much that it is difficult for me to take an unemotional standpoint, and I daresay this is true of most people . . .' On the other hand he found the reciprocal bombing of the civil population peculiarly senseless, and he could never envisage a British army expelling the Germans from their European conquests:

> I do admire immensely not only your intellectual analysis but your moral courage in making it. You may be right, and in fact I am not in a position to dispute your arguments. The large part of me which loathes all suffering and, indeed, all military activity, cries out to agree with you. What (if one may suggest it) you do not seem to have taken fully into account is the effect on:
>
> (a) Hitler, who might be mightily encouraged;
> (b) The United States, who just love us to fight their wars for them;
> (c) The people at home, who might feel they have been betrayed.
>
> Would the United States go on producing arms for us?
> Would Hitler give up any of his gains?
>
> On the other hand, I see clearly that these questions remain just as urgent if things go on as they are, and your solution would give us the inestimable benefit of even an uneasy peace.[22]

In a long and important reflection entitled 'The Reckoning' dated 17 November 1940, Liddell Hart pursued the argument that even if Britain could achieve eventual victory the resulting conditions might hardly be better than peace terms negotiated now from a position of inferiority. The devastation would be greater than in any indecisive war which ended earlier, while financially Britain was likely to become the poor dependant of America—if we did not become 'impoverished Communists'.

> Worst of all the probable consequences is that in trying to root out Hitlerism we shall sow a fresh crop of Hitlers, not only in Europe but in our own soil. Just as the last 'war for democracy' caused widespread growth of a more drastic autocracy, so this avowedly

totalitarian struggle must naturally tend towards total bureaucracy under despotic guidance. The longer it endures the more firmly will the chains be rivetted. (Neither the state of freedom nor the sense of freedom which we knew before the last war was ever recovered after it).

The prospect is all the darker because the means to which we are now reduced in pursuing victory—the sustained bombing of industrial targets combined with a blockade aimed at the gradual starvation of all Europe under the enemy's control—tend to have a more degrading effect of [*sic*] civilised life than what has been experienced in any previous modern wars. (The theory of *blitzkrieg* is humane by comparison, though it lends itself to aggressive purposes—while it cannot be effectively applied by the opponents of aggression).

It is regrettable, to say the least, that those who have gone to the rescue of humanity from Nazi tyranny, should be led to inflict such limitless misery on millions of innocent human beings in the occupied countries as well as in Germany, in trying to free them from the rule of a limited group. And this time our own people are themselves equally exposed to such effects—from the counter-bombing and counter-blockade of their more vulnerable island ...

Since the collapse of France, and the end of the land campaign, this war has assumed the character of a giant's game of skittles. The longer it continues the more likely that the match may be extended to farcical dimensions—with one side chiefly using missiles produced across the Atlantic to throw at Germany's cities, while the other makes them in the remoter parts of Europe, to knock down England's cities. It is a grim outlook for all those placed in the skittle-alley. And we should be wise to take account, now, of the ultimate effects of the Americans' growing ardour to beat the Nazis—lest it may lead to them becoming determined 'to fight to the last English cottage'.[23]

The idea that even if victory were possible a new 'Hitler myth' might be created was developed in Liddell Hart's next memorandum:

Even if we could eventually succeed in destroying Hitler's 'New Order' by force, the resulting state of chaos and misery might produce the belief that a heaven-sent solution of Europe's troubles had been destroyed by mere force. Thus out of the ashes of the Nazi regime a Hitlerian legend might arise—to endanger the future even more than did the Napoleonic legend. That would be the supreme irony of our attempt to crush force by force.

In *Why Don't We Learn From History?* this idea was taken still further:

> If post-war conditions should prove disappointing, the experience of history suggests the possibility that he [Hitler] may be pictured as the man who might have led the European masses into 'a land flowing with milk and honey' had he not been first thwarted and then overthrown by jealous opponents and vested interests![24]

This passage does not suggest that Liddell Hart possessed much knowledge of conditions in Nazi-occupied Europe. Nor does he seem to have grasped the effect of the military triumphs of 1939–40 in strengthening the Nazi Government's hold over the German people.

At the end of 1940, then, Liddell Hart saw no meaningful ending of the war except by a negotiated peace. He favoured land operations against Italy, but could not see how Britain could defeat Germany 'on orthodox lines', i.e. by invading the Continent. As regards the indirect means available to Britain, he was bitterly opposed to strategic bombing and had severe reservations about the morality of blockade. The latter, he noted unrealistically, ought to be adapted so as to pinch Germany rather than the peoples in her grip; otherwise by starving the peoples of the occupied countries Britain might commit 'a greater crime against humanity and civilisation than anything that even a tyrannical dictator has contemplated'.

In lieu of military strength Britain should employ moral and psychological weapons. At the conclusion of *Dynamic Defence* Liddell Hart launched an eloquent plea for 'a constructive and dynamic peace plan that will stir the imagination of our people and the peoples of the world. It must appeal to the individual everywhere—in the enemy countries as well as in those that have been subjugated.' Also, since the totalitarian regimes offer the prospect of material benefits, we must stress the fundamental difference that separates our conception of civilized life from theirs—the idea of respecting individual freedom.[25]

These were no doubt admirable sentiments, but what were the chances of their having any practical effect? The difficulties were many and formidable. Britain's much-vaunted freedom had not stood her in good stead in the later 1930s. Could she afford to increase the liberties of her citizens at the very time when she was fighting for her existence after a series of humiliating defeats against a supremely confident adversary? Could she effectively transmit these ideas of freedom to the subjugated European peoples, and if so would such as the Czechs, Poles and Norwegians be impressed in view of Britain's inability to save them from being overrun in the first place? Above all, would they be impressed by British propaganda—however well founded—unless Britain herself was seen to be going all out first to defeat invasion and then to lead the counter-attack against the Axis Powers?

In the early months of 1941 Liddell Hart's main theme continued to be the need to prevent a German victory, while at the same time deploring the two-pronged instrument of blockade and strategic bombing which the Government was using to achieve it. As regards grand strategy, although there are several references to an Anglo-American partnership, he does not seem to have taken much interest personally in persuading the United States to enter the war. Thus a typical reflection on 12 April 1941 reads:

> If we can prevent a German victory, and also avoid bleeding ourselves to death, we may be able eventually to create, in conjunction with America, a combination of air and naval force with economic power that will be capable of turning the tables—*even without the use of force* [my italics]—and determining the future world order.[26]

In few aspects of the war did Liddell Hart's dispassionate detachment stand him in better stead than in regard to the Soviet Union. After Munich he had urged that an agreement with her was essential if an effective diplomatic barrier was to be raised against Hitler. He reacted with restraint when the Soviet Union attacked Finland and consistently warned against any action which would entail the risk of war with her. Indeed from early in the war, though not optimistic, he saw the possibility of the 'two thieves' falling out as one means of ensuring Britain's survival.

With the launching of Operation Barbarossa, Liddell Hart at once recognized that the whole face of the war had changed both militarily and ideologically. Now, for the first time, Britain was presented with the possibility of military victory, as distinct from the frustration of a German victory. He correctly judged that Hitler was taking an enormous gamble in aiming to knock out the Soviet Union by the *blitzkrieg* methods which had worked elsewhere before throwing his whole might against Britain. Without overtly committing himself as to whether the Soviet Union could survive, he urged that Britain should do as much as possible to distract Hitler by increasing the air offensive against his war potential to the maximum and exploiting any opportunity for sea-borne raids.

It was not Liddell Hart's way to admit that he had been wrong in his previous pessimistic views and that Churchill's optimism had been justified; he merely remarked that hitherto the optimists' belief in eventual victory had been supported only by instinctive faith: 'the war held no possibility of victory for us until Hitler committed this fatal mistake'. His fundamental opposition to a fight to the finish was unaffected by Hitler's folly. Nevertheless, so far as British politics was concerned, Russia's entry marked the great divide. Hitherto there had been an uneasy partnership between those willing to fight to the death to preserve Britain's independence but pessimistic about the possibility of a Continental crusade, and the Churchillians. Churchill's great gamble, taken at long odds in May 1940, proved a winner on 22 June 1941.

After carefully following the course of the German advance through the summer, by September 1941 Liddell Hart was cautiously optimistic that the Russians would hold out, but he warned 'We have a long and hard road ahead of us. It will not be made easier by paving it with "gold bricks".' He avoided both the extremes of those who expected the Red Armies to be beaten within a few weeks, and those who began to write off the Germans as soon as they suffered checks in the early autumn. By January 1942 he was sufficiently confident of the Soviet Union's survival to ask 'shall we be capable of invading the Continent in 1942, or later? And will it be necessary?' Since he did not apparently take into account America's possible involvement in the European theatre, he answered all these questions in the negative. Since we lacked a 'New Model' army and modern landing craft, 'our hopes must rest with the leaders of Germany and their capacity to destroy the strength and morale of the German Army in vain attack—in unavailing pursuit of the victory that they deemed so certain in 1940'.[27] This Hitler effectively did by his offensive strategy on the Eastern Front in 1942 and 1943.

After June 1941 the largest missing piece in Liddell Hart's strategic jigsaw puzzle was America's participation in the European theatre. Whether or precisely when America would have declared war on Germany can never be known, since Hitler gratuitously declared war on her after the Japanese attack on Pearl Harbor.

Liddell Hart's initial reaction to the latter event was that anyone with a historical sense should have foreseen it. To his credit, he quickly perceived that even while the Japanese forces were still advancing at an astonishing pace—and with very light losses—they were giving hostages to fortune by spreading their efforts so widely and disembarking such large military forces on islands where they could be isolated by superior sea and air power. He was on more dubious ground in attributing the Allies' defeats in the Pacific primarily to their propensity to attack.[28] From his writings generally in 1942 it seems that Liddell Hart was slow to grasp the vital part which American forces would play in the European theatre. A possible explanation is that he was well aware of America's military weakness in the short run and still hoped for a compromise peace before her vast war potential had been developed.

Proof that the Soviet Union's enforced entry into the war against Germany did not alter Liddell Hart's fundamental thinking is to be found in his important memorandum 'The Eternal Will O' The Wisp in War' dated 3 September 1941, copies of which were sent to Lloyd George, Lord Noel-Buxton, L. P. Jacks, Kingsley Martin, Hore-Belisha, James Maxton, Stephen King-Hall, Gilbert Murray, John Brophy and others. This memorandum urged the need for a compromise peace for the reasons outlined earlier, as implied by the title, and its theme was that 'victory over another great power never fulfils expectations of a good and lasting peace'.[29] Hore-Belisha disagreed flatly with his former military mentor, but John Brophy's reply was the most thoughtful and is worth quoting at length, not only because it provoked a dialogue with Liddell Hart but also because it exemplifies the realist's and the idealist's differing views of the nature of the war.

> I see your point of view very clearly [Brophy replied] but I am not sure that it is not too long-term for our present necessities. What I mean is that if there is no real prospect of our gaining an outright victory we surely need not worry too much about the after-effects of such a victory. In the meantime, we are up against an enemy of a sort we have never encountered before, and your suggestion to force him to make peace on our terms rather begs the question of how he is to be made to keep the agreement. This is how it seems to me. No doubt we shall not get the sort of victory the Government is preaching, but the very nature of the Nazis seems to me to preclude any compromise. There I am with Churchill and not with you.
>
> Again, another objection. What you say about Churchill, is very illuminating, and as far as I can judge justified. But what alternative have we for Prime Minister? And have you not ... failed to give him credit for the enormous moral transformation he has brought about in this country? ... Churchill has lifted us out of the depressed era of Baldwin, Ramsay MacDonald and Chamberlain, with the result that for the first time for many years I feel the air in this country is more or less fit to breathe.[30]

This drew forth another reply from Liddell Hart in the form of a memorandum which was also widely circulated. Three related points in it are worth mentioning. Firstly, Liddell Hart thought his critic was falling for the old unhistorical assumption 'that we are up against an enemy fundamentally different from any we have encountered before'. The same error had been committed when we fought Louis XIV and Napoleon. Secondly, by our tentative moves we had repeatedly compelled the Germans to forestall us, thus causing the war to spread and become unlimited. Thirdly, our present leaders had forgotten the fundamental condition which produced our traditional strategy and policy, namely pursuing in war the aim of an advantageous peace by negotiation. Step by step we were being led into the position where we might not have enough bargaining power, and might be faced with total defeat—totally unnecessary with a wiser policy.

To this Brophy rejoined:

> What I am questioning is not so much the validity of your arguments as their practical application to popular opinion today. I should imagine it would be very difficult to make people enthusiastically for anything less than 'Victory', and there is in many people's

> minds a long founded and, on the whole, justifiable suspicion of compromises on the part of their leaders ... Also only by going all out can we keep unity with the United States, the Soviet Union and the European peoples.

Liddell Hart explained that by 'negotiated peace' he envisaged a situation where our combined invincibility and 'nuisance value' would compel Hitler to concede the release of most of the territory he had conquered.

> Any other kind of ending, short of our defeat, would not be negotiated—though it might take the form of a suspension of open hostilities growing out of a mutual recognition that the probable result of pursuing the struggle actively would be mutually devastating without compensating gain.[31]

For a brief period at the beginning of 1942 it looked as though Liddell Hart might be offered an official appointment as a military adviser, but in the end nothing came of it. In February and March his diary entries show that he spent a good deal of time advising the new Secretary of State for War, Sir James Grigg, particularly on senior Army appointments. On 18 March Grigg's Parliamentary Private Secretary, Sir Archibald Rowlands, who was an old friend of Liddell Hart's, suggested the creation of a body of scientific experts under Sir Henry Tizard and Liddell Hart. They could take over a mansion near London. Rowlands felt that the Army, like the R.A.F., should have an operational research section at each of its major headquarters. What the Army needed was 'twenty bits of L.H.' divided among the various expeditionary forces. There was more talk about this 'Brains Trust' at the end of April, but by then Grigg's enthusiasm was fading in the face of the soldiers' objections. Liddell Hart also heard that some senior generals, particularly Lieutenant-Generals Pownall and Adam, had opposed his appointment as adviser to Grigg's predecessor, David Margesson.[32]

Another old friend, Ivan Maisky, the Soviet Ambassador in London, was astonished to learn that Liddell Hart's expertise was not being used by the Government. According to Lloyd George, Maisky had told him that 'We regard Liddell Hart as the best military brain in England, and the only one'.[33]

Liddell Hart could surely have played a valuable part on the level of Army organization, training and operational planning, and it is worth emphasizing that he remained intensely interested in such practical matters. But on wider issues of policy and strategy his views were so much at odds with Churchill's that he would have had to abandon his basic tenets in order to survive. At the very time, for example, when Rowlands was trying to set up the 'Brains Trust' under Tizard and Liddell Hart, the latter was writing to Edward Woods, Bishop of Lichfield, in very critical terms about Churchill. To the Bishop's description of the Prime Minister as 'our heaven-sent leader', Liddell Hart retorted that in the last year before the war he had come to feel that Churchill's power for evil was potentially greater than his power for good—'that his almost unique egocentricity, his dramatic sense, his lack of scruple, and his lack of judgement, made a combination that might be destined to lead this country to disaster'. Events since then had tended to confirm these fears. Such heterodoxy would hardly have been welcome in the Prime Minister's entourage in 1942.[34]

Indeed Liddell Hart's sturdy independence of mind put him permanently out of step with public opinion. He was always temperamentally inclined to make himself the spokesman of unpopular, minority views and whether he was right or wrong on particular issues, he was performing a valuable democratic role. In the period of Britain's isolation he had been unable to find any logical reasons to believe in ultimate victory; now with the United States and the Soviet Union in the war he began to urge other arguments in favour of a speedy negotiated peace. These apprehensions are evident as early as 19 March 1942 in a talk with Lord Beaverbrook, who was seeking his views just before going to Washington. In discussing the possibility of a second front in 1942 Beaverbrook expressed doubts about whether we could win the war and cited Fuller as a military expert who would not commit himself. Liddell Hart pointed out the folly of our absolute victory cry, contrasting it with Stalin's strategic wisdom in showing the German people a way out, thus loosening their resistance. Beaverbrook agreed but pointed out the difficulty in a democracy of expressing the war aim in anything but absolute terms.

Liddell Hart suggested that Washington was bound to become the centre of the Allied war effort, which would place Britain in an awkward position at the end of the war if we were clearly poor dependants of the United States. The only hope he could see of Britain retaining her dominant influence was an immediate merger with the Dominions in a greater Commonwealth.[35]

On Beaverbrook's return to London he told Liddell Hart of the extraordinary wave of optimism he found sweeping the country and the City. It was clearly based on events on the Eastern Front but he could not see any justification for it. Liddell Hart agreed that it was far too soon to write off the German campaign in Russia, and he later advised Beaverbrook that it would be madness to attempt to open a Second Front in 1942.[36]

Nowhere was Liddell Hart's opposition to Churchillian strategy more pronounced than in his abhorrence of the strategic bombing of cities. It was this theme, more than any other, which formed the basis of Liddell Hart's close association with George Bell, Bishop of Chichester, who probably forfeited his chance of becoming Archbishop of Canterbury by his brave and forthright criticism of Government policy.

As early as May 1941 Air Marshal Sir Richard Peck, Assistant Chief of the Air Staff, told Liddell Hart the 'inside story' of the British initiation of strategic bombing in the summer of 1940, adding that it was mainly due to Churchill and Air Marshal Sir Cyril Newall, the Chief of Air Staff. In December 1941 an unnamed R.A.F. officer informed Liddell Hart (correctly) that, official reports notwithstanding, the bombing offensive was failing badly; the recent big daylight raid on Cologne, for example, had been a disaster. Bishop Bell accepted Liddell Hart's evidence that technically Britain had started the strategic bombing of cities, but suggested by way of mitigation that Hitler intended to invade England and would surely have bombed London whether or not we had bombed Berlin. He pertinently cited the 'complete ruthlessness' with which the Luftwaffe had been employed in Poland, Denmark, Norway and the Low Countries. (He might have added the destruction of Guernica on 26 April 1937.) Liddell Hart countered with one of his basic beliefs: the Germans employed military force for definite ends (as in their limited use of trench raiding in the First World War), and would *not* have bombed London *before* attempting an invasion had we not bombed Berlin. British policy was foolish on practical as well as moral grounds, since we stood to suffer more than the Germans in an all-out bombing contest.[37]

Liddell Hart's attitude to strategic bombing is conveniently summarized in a Reflection arising from the Cologne raids which he wrote in June 1942;

> It will be ironical if the defenders of civilisation depend for victory upon the most barbaric, and unskilled, way of winning a war that the modern world has seen.
>
> In the first two years of the war the Germans owed their victories not merely to superior force but to the fact that they brought the art of war to a new pitch of skill. We are now counting for victory on success in the way of degrading it to a new low level—as represented by indiscriminate (night) bombing and indiscriminate starvation.
>
> That will at least be a natural continuation from the last war, where it was the Germans who showed the only skill of generalship that was seen in Europe, yet the Allies who wore them down by sheer attrition—with shells on the battlefield and hunger in the home.
>
> If our pounding of German cities, by massed night bombing, proves the decisive factor, it should be a sobering thought that but for Hitler's folly in tackling Russia (and consequently using up his bomber force there, as well as diverting his resources mainly into other weapons), we *and* the Germans would now be 'Cologning' each other's cities—with the advantage on Germany's side, in this mad competition in mutual devastation.
>
> Such *would have been* the consequences of our chosen method of winning the war—the method which a large section of opinion was itching to unloose during the first winter of the war, and eventually did unloose, prematurely, in the summer of 1940—so that for a year and more, until Hitler turned East, we 'took punishment' out of all proportion to what we were able to inflict.

He denounced the bombing campaign in similar terms in *This Expanding War* (published in the autumn of 1942) but included a newspaper article dated 24 June 1941 in which he called for 'the maximum development of our present air offensive on Germany's war potential'.[38] Since Bomber Command was at that time unable to pin-point military targets, it was really a stark choice between doing as well as possible in adverse conditions or giving up the bomber offensive altogether.

In July 1943 General Sir Frederick 'Tim' Pile, who commanded Britain's anti-aircraft defences throughout the war, remarked, according to Liddell Hart:

> Winston is pinning all his faith to the bombing offensive now. The devastation it causes suits his temperament, and he would be disappointed at a less destructive ending to the war. Having seen him at close quarters on many occasions, particularly in the Battle of Britain, Pile feels that he is a 'masochist'—danger and destruction are the breath of life to him.

Liddell Hart's temperament was very different indeed. He believed that it was foolish to combine strategic bombing with a policy of unconditional surrender, but he decided against sending Churchill a memorandum on the subject 'since his mind had such a destructive cast that it might hardly be penetrable by such a different idea'. Air Marshal Peck agreed with Liddell Hart on this point; in his view also, Churchill was 'not a good psychological strategist'.[39]

Liddell Hart's opposition to strategic bombing remained substantially unchanged to the end of the war. In April 1944 a war reporter, Alexander Clifford, reinforced Liddell Hart's critical views by retailing impressions of an officer (George Miller) who had recently escaped from Germany. Miller apparently felt that our bombing was leading the Germans to the view that we were more barbarous than the Russians. Military morale was still high but civilian morale was declining. According to Clifford, on his recent tour of the Mediterranean theatre he had heard that the Americans had casually bombed Innsbruck not for any military reason but simply because it was a convenient place on the map; while Air Marshal Coningham was alleged to have advocated the bombing of Venice 'to show the Italians we mean business'. In April 1945 Liddell Hart noted 'Our strategic bombing plan had been a failure . . . Our real success was in bombing oil refineries and synthetic plants. If we had concentrated on this we should have done much better.'[40] Subsequent research shows that there was a good deal of justification for this criticism, particularly in the last eighteen months of the war, but it overlooked the counter-argument that however ineffective the early phases of strategic bombing had been in material terms, they had provided experience on which success was eventually built. If Bomber Command had been able to locate and hit military targets with reasonable accuracy throughout the war, presumably Liddell Hart's moral objections would also have disappeared.

By the beginning of 1943, with the success of the Anglo-American operations in North Africa and with the Russian stranglehold tightening around Stalingrad it was clear that, provided the Alliance remained firm, military victory in Europe must be only a matter of time. In accordance with these developments a new note is heard in Liddell Hart's writings, epitomized by the heading of a reflection on 10 June 'We have won the war—but don't know it: We are now stupidly fighting on to sacrifice life, wealth, health and independence—in short to lose the peace'. His ideas were set forth in more detail in a paper entitled 'The Outlook—Spring 1943' in which he wrote:

> Victory is a *possibility*; stalemate a probability; loss of our purpose in fighting almost a certainty. Peace through victory in a year or two's time would probably mean that we should lose more, in every sense, than by peace through stalemate this year. Victory could only come this year if Russia were to succeed in winning it 'off her own bat'.
>
> At this moment Hitler's power of aggression, probably even his desire, have been quenched more effectively than Louis XIV's was after the Peace of Utrecht—the best peace ever achieved in modern times...[41]

It was entirely consistent with Liddell Hart's general outlook on the war that he should regard President Roosevelt's unpremeditated declaration of the 'Unconditional Surrender' policy at the conclusion of the Casablanca Conference in January 1943 as an unmitigated disaster. Even from a strategic point of view, he wrote to Bishop Bell, the declaration seemed to him 'the most stupid and untimely step that could have been taken—and the best possible reinforcement to Hitler'. The Bishop deplored its application to Italy, and agreed that it would be exploited for propaganda purposes by Goebbels.[42]

Liddell Hart was stimulated to see a deeper connection between strategic bombing and the unconditional surrender policy by a letter from his brother-in-law Barry Sullivan, then serving in the R.A.F. in Malta. The latter, depressed by the indiscriminate destruction of strategic bombing, felt that the Government 'has become Frankenstein, dominated by its own creation—the monster of Bomber Command'. Yet on re-reading *The British Way in Warfare* it seemed to him that strategic bombing was the 'natural offspring' of British historical strategy, which in the past had been more effective and more economical of life than the Continental way of war. How would Liddell Hart resolve this apparent contradiction? Liddell Hart replied as follows:

> Any dispassionate historian and student of war would have to admit that our traditional strategy, which we followed during the 16th, 17th and 18th centuries, was *inherently* more 'barbarous' than the strategy of the Continental tradition and of Clausewitz's theory—because it aimed at the will of the opposing people, rather than at their main armed forces, and was thus in a sense striking at the non-combatant population. At the same time, it proved in practice less damaging, and more reasonable.
>
> The explanation of this practical paradox lies in the reasonableness of our policy regarding the object. In the past, we were usually willing to accept a negotiated basis of peace when the enemy had become sick of the war, and was willing to climb down—to abandon his opposing policy. We did not pursue the fight to a finish—which is apt to entail not merely the exhaustion of the aggressive impulse, but the mutual exhaustion of the capacity to rebuild peace.
>
> It is the combination of an *unlimited* object with an *unlimited* target of strategy—the combination of a demand for the other side's unconditional surrender with a strategy on our part of total blockade and air bombardment—that inevitably makes our strategy a reversion to 'barbarism', and thus a spreading menace to the relatively shallow foundations of civilised life.

In a further reflection he showed how the policy of unconditional surrender would rule out the two possible ways in which strategic bombing might have achieved its object, i.e. the enemy's capitulation, either by a revolt which overthrew the regimes in Germany and Italy, or by popular pressure for peace compelling the regime to capitulate. He concluded that:

> Insistence on 'unconditional surrender' thus aids the hostile regime in keeping control of its people, and convincing them that they have

> no alternative than to sink or swim with the regime. The effect tends to be like that of a frightened crowd pressing down a passage towards a barred gate—more effective in suffocating the leading ranks of the crowd than in breaking open the gate. Beyond this is the question whether people who feel themselves the target of an unlimited attack with an unlimited object—i.e., an unconditional surrender that provides no safeguards against their maltreatment when submission has rendered them completely helpless—will not be inclined to rally to the regime, tyranny though it is, which at least organises their defence. In such an impasse, the failure to make clear any limitation of the object may blunt the edge of the bombing weapon.[43]

This discussion of the Allies' unconditional surrender policy raises the large question of Liddell Hart's interpretation of Hitler's war aims and his rationality in pursuing them. It raises also the related problem of the anti-Nazi opposition inside Germany and its attitude to peace terms should Hitler and his henchmen be deposed. In general it can be said that Liddell Hart took a very optimistic view of Hitler, picturing him as a rational statesman with limited aims who would see the folly of persisting in war once it had been demonstrated that those aims were not to be obtained by force. Perhaps some of Lloyd George's admiration for the Führer had rubbed off on him. In the memorandum quoted above, for example, Liddell Hart wrote that, immensely helped by Russia, we had 'succeeded in crippling the enemy's power of aggression, and making him anxious for peace on the terms we defined on entering the war, or on any terms, indeed, that did not involve his complete downfall'. Was there even a shred of evidence for believing that this represented Hitler's attitude in 1943 or after? Liddell Hart displayed remarkable detachment in writing of 'Hitlerism'. 'To blame Hitler for the war', he reflected in November 1942, 'is to exaggerate both his guilt and his importance. Hitler and the war are but episodes in a Political-Economic Revolution that is in progress in our time. They are eruptions on the skin.'[44]

One of Liddell Hart's fundamental assumptions, which he shared with Bishop Bell, was that the Allies should encourage the German and Italian people to throw off their tyranny. Thus Liddell Hart approved of the Bishop's motion in the House of Lords on 10 March 1943 calling for the British Government to distinguish between Hitlerites and non-Hitlerites in stating its policy for a post-war Germany. Liddell Hart was also inclined to give the benefit of the doubt to rumours of a German opposition to the Nazis. In August 1942, for example, he learnt from David Owen, Sir Stafford Cripps' Personal Assistant, that the anti-Nazis had put out peace feelers through contacts in Sweden: 'None had promised anything better than the situation at the time of Munich and there had not been any for several months.' Church and Army leaders were said to be prepared to carry out a *coup d'état* if they could get adequate assurance of our subsequent attitude. Liddell Hart felt the British response was too unhelpful, whereas Owen was sceptical about the authenticity of feelers purportedly coming from German Church leaders. In September 1942 Bishop Bell told Liddell Hart of personal contacts he had made in Sweden with members of the German opposition; he had passed on the information to Cripps and Eden, but the latter would make no move until the anti-Nazi forces had shown their capacity for action. The Bishop had heard in Sweden that the German opposition was handicapped by the Allies' refusal to make any distinction between good and bad Germans. In an extremely interesting letter dated 22 March 1943 Liddell Hart wrote to Bishop Bell that he had heard via Swedish contacts that the Germans had tried to negotiate peace with Russia. It was said that von Paulus (commanding the beleaguered Sixth Army in Stalingrad) had been promoted Field Marshal to give him the necessary status to negotiate. The basis suggested was the evacuation of all Russian territory, certain territorial compensation to Russia at the expense of Finland and Rumania, and a measure of reparation for damage done in the invasion.[45] This was almost certainly the source of Liddell Hart's references during the war to his belief in Germany's willingness to negotiate peace in 1943.

While Liddell Hart cannot be blamed for ignorance of the extent and strength of anti-Nazi activity inside wartime Germany, he was guilty of wishful thinking in placing reliance on the dubious rumours emanating from Sweden. The German opposition groups were ill-organized and disunited; some of the more prominent members like Carl Goerdeler wished to retain a large part of the Nazis' conquests; and no opposition group, had it been able to take power before Germany's complete defeat, could have made the kind of self-abasing peace settlement which Liddell Hart envisaged. But the important fact remains that the opposition never came close to overthrowing the Nazi regime, not even after the plot of 20 July 1944. The Allied Governments had to deal with Hitler, and their jaundiced view of that individual were much nearer reality than Liddell Hart's. So far from promoting von Paulus in order to negotiate peace, it is more likely that Hitler expected it to stiffen his resistance and determination to avoid capture. Hitler's whole conduct of the war on the Eastern Front flew in the face of reason and the more the tide turned against him the more irrational he became.[46] Much of this evidence was not available to Liddell Hart during the war, but even the course of military events should have caused him to question the assumption that Hitler was a statesman pursuing reasonable goals in a reasonable way. Instead he seems to have allowed too much weight to his idiosyncratic view that the Allies were largely to blame for provoking Hitler to successive conquests and for making the war 'total' by their policies of blockade, strategic bombing and unconditional surrender.

Liddell Hart's mental detachment and non-involvement in the conduct of the war enabled him to take a dispassionate and at the same time remarkably prophetic view of the post-war European world. In his admirable memorandum on 'The Future Balance of Europe' dated 1 October 1943, he saw very clearly that Russia would replace Germany as the dominant power in Europe and that in the long run this might be even more dangerous than German hegemony. Russia obviously intended to absorb the Baltic states and a large part of Poland at the least.

> The immediate consequences of victory would probably be the Red Army's occupation of the whole of Central Europe and a large part of Germany. Russia alone would have the strength to place an effective army of occupation in these countries. At the same time the Anglo-American armies may occupy the countries of southern Europe and some part of Germany.

He foresaw in general terms the crisis situation (that actually arose at the time of the Berlin airlift) which would occur 'if Stalin were to serve notice to quit upon the Anglo-American occupying forces'. In any event Britain would find herself in the awkward position of a buffer state between the great powers, but on the wrong side of the Atlantic. The only state which could really add to the strength of the buffer (i.e. Germany) was the one we were now aiming to smash. 'The irony of the situation is that the fulfilment of our aim of complete victory is bound to destroy the only possible breakwater of any real strength.'[47]

Field Marshal Smuts also saw that the defeat of Germany would only put a new colossus in her place but, according to Liddell Hart's notes, he failed to bring out that a German army would be essential to any post-war European group designed to offset Russian power.[48]

Of those who commented on Liddell Hart's memorandum, Oliver Harvey replied tersely that he did not agree; Geoffrey Faber thought that Liddell Hart might be right but the tide would probably be too strong for him; Lord Geddes agreed that the unconditional surrender policy was an error; and Gilbert Murray, while agreeing with Liddell Hart, thought that the Government's policy was more moderate than its propaganda. John Brophy's was again the most interesting reply. He was more optimistic about the future policy of Russia on two counts: firstly, there was a good chance of her becoming Westernized and liberalized in outlook; and, secondly, she would be so preoccupied with reconstruction that she would have no time for imperialist adventures. Liddell Hart replied that he had a strong hunch that Russia would be troublesome after the war and the main danger would come from American–Russian friction. This was not so obvious in 1943 as it seems now. In March 1944 Cripps could still assert that there was no likelihood of a clash between Russia and America after the war.[49]

This memorandum and the correspondence that followed it demonstrated Liddell Hart's capacity for brilliant insights in the field of grand strategy. In 'The Future Balance of Europe' and in reflections on 'War Aims' and 'Victory' which followed, he outlined a position that remained substantially unchanged for the rest of the war.[50]

In March 1944 a bizarre episode occurred when Liddell Hart was suddenly summoned to attend the War Cabinet. Was this a belated invitation to place his expertise at the government's disposal? Far from it, for on arrival he was sternly cross-examined by Sir Hastings Ismay and Ian Jacob of the War

Cabinet Secretariat on how he had obtained advance notice that the invasion of Europe would take place in Normandy. Liddell Hart had considerable difficulty in persuading his interrogators that a landing in Normandy, which he had mentioned in a memorandum dated 25 January, was simply the obvious deduction which any competent strategist would make from the vast build-up of Allied forces in the south-west of England. He eventually convinced them that there had been no breach of security. There was an amusing sequel. Once he had been cleared, Liddell Hart suggested to Jacob that his name be used as a red herring: his appointment as a Government adviser should be announced so that the Germans would deduce, incorrectly, that an 'indirect approach' in the Bay of Biscay was being planned. In regretfully informing Liddell Hart of the War Cabinet's rejection of this idea, Jacob mentioned that he had found his recently published book *Thoughts on War* very stimulating, but pointed out that Liddell Hart enjoyed the privilege of the writer over the man of action: others besides himself could *see* an ideal solution but they often had to compromise to meet the needs of allies or to accommodate other pressures.[51]

One of Liddell Hart's outstanding traits, which this chapter has tried to bring out, was his attempt to keep a calm, fair-minded attitude towards Germany and the Germans in face of the virulent propaganda and emotionalism which swept Britain as a result of the blitz and subsequent events. In attempting to combat the anti-German feelings, at times approaching hysteria, of his own countrymen there was a danger that Liddell Hart might carry fair-mindedness to the point of self-delusion. While Liddell Hart's refusal to hate the German people was admirable, there is a suspicion that he was either ignorant of, or unwilling to believe, the barbarous and bestial nature of Nazi practice within Germany and throughout occupied Europe.

This suspicion may be illustrated from a fascinating correspondence with John Brophy in the summer of 1944 initiated by Brophy's remarking that, if practicable, at the end of the war he would like to see all the Germans wiped out.[52] Since Liddell Hart regarded Brophy as mild and broad-minded in his opinions, he asked whether this outburst of anti-German hatred (worse than Vansittart's) was due to nerve strain resulting from the flying-bombs. In pleading for a tolerant view towards Germany, he pointed to what we were doing to their cities.

Brophy suggested in return (and it is a thought which occurred independently to the present writer) that in making strictures on British conduct of war he was thinking too exclusively of the Anglo-German contest. 'How did the Germans behave in Poland, in Russia, in France and so on?' wrote Brophy

> What I feel is that unconsciously you have taken to applying two standards to Britain and Germany. A few black specks show up on the British record, a few white ones on the German. But taken as a whole the British is morally clean, and the German filthy. The concentration camps alone are enough to indict the Nazis. I stick to the general proportion and I see the British with all their faults as the most civilised and innocent nation in the world...

Liddell Hart's reply took up Brophy's points one by one, and when he came to the reference to concentration camps he seemed to discount the stories of German atrocities as Polish and Russian propaganda.[53] While he agreed that the *pre-war* Nazi concentration camps provided abundant evidence for indictment, even so a number of friends had told him after release in 1940, after being interned in Britain as aliens, that they had found the Nazi camps preferable to ours. The most reliable evidence we had so far, he pointed out, was from the liberated areas, including the Channel Isles, and in these it was to be noted that 'the behaviour of the Germans was much more correct and restrained than we had imagined'.

Liddell Hart's tendency to magnify Britain's errors while minimizing Germany's can be accounted for in two ways: in practical terms he was personally acquainted with his own country's faults and privy to continual inside gossip; in theoretical terms he was prepared to push rather far the notion that the authoritarian aggressor has limited aims and strict rules of conduct while the peace-loving democratic victim, once provoked into action, wages unlimited war both as regards conduct and political objectives. If the second part of the latter assumption was largely true of the democracies in the Second World War, it did not follow that the first was also.

This line of reasoning may be further illustrated by a reflection on the Codes of War stimulated by the information that the British had been ordered to bind prisoners' hands in the Dieppe raid:

> Likewise, on a higher plane, our methods of war tend—partly by force of circumstances—to be more inhuman than those of battle-minded Continental countries. In the present war that tendency has been accentuated—the dual plan of seeking victory by bombing cities and by starving all Europe is, fundamentally, the most barbarous form of war that the modern world has seen.

In the political sphere Britain was on a higher level of decency and humanity.

> But in the sphere of war our policy tends to be less human than that of the Germans and would seem to be so in practice—their inhumanity being manifested rather in the political consolidation of military results.[54]

In fairness to Liddell Hart it may be that in trying to combat 'the Illusion that the latest enemy is "Different"', he genuinely came to believe that Nazi Germany was no more evil than the France of Louis XIV or Napoleon. It is certainly curious, in view of his complete integrity in preserving evidence, that he does not seem to have received information either about the concentration camps or of particular German military atrocities during the war.[55] In September 1941 he referred to the hatred which 'the Nazi gospel of hate has inspired' without referring to Nazi practice.[56] One gathers also that he shared Sir Hartley Shawcross's amazement in September 1945 to learn that a Nazi had mentioned that four million Jews had been killed.[57] One's suspicion that Liddell Hart really was profoundly ignorant of the nature of the Nazi regime is strengthened by his remarkable reaction to an article by Arthur Koestler entitled 'The Mixed Transport' in the October 1943 issue of *Horizon*.[58] This grim account purported to describe a journey in Eastern Europe of seventeen closed cattle trucks carrying Jews to the labour camps or to execution. Liddell Hart significantly entitled his reflection 'A Fall from Reality'. He went through the article pointing out 'improbabilities and absurdities' in detail, such as whether the carriages were opened *en route*, whether or not the passengers could see out, and proved to his own satisfaction that the story was false. Unfortunately, as we know, Koestler was absolutely right; indeed he had only lifted a tiny portion of the veil covering the horrors that were taking place over a vast area of Nazi- (and Soviet-) occupied Europe.

There are no new strands in Liddell Hart's thinking in the last year of the war. If possible, the tone of his memoranda and reflections becomes even gloomier as he sees his predictions in 'The Future Balance of Europe' being fulfilled: the Allies persisting in their policy of unconditional surrender with the effect of reducing Britain to a second-grade power economically dependent on the United States, replacing German dominance of Europe by Russia and sowing the seeds of a Hitler myth. His strictures on the barren results of victory by the complete conquest of Germany were true enough:

> We have a ruined Europe to deal with, besides being more exhausted ourselves; *and* have lost our honour over the pledge to Poland. Europe is divided into two potentially rival spheres, without any buffer between them—owing to the destruction of Germany's defensive power. Not only Poland, but all the other countries of Eastern and Central Europe are enclosed in the Russian sphere, with puppet governments that are under Russia's firm control. On the other side of the hill to that solid *bloc*, where dissenting elements have been purged, lies an Anglo-American sphere which is seriously weakened by internal dissensions as well as by devastation. Already in France and Italy a violent swing towards 'Communism' is taking place, as hunger and trouble multiply. Stalin can afford to smile, for he stands to win both ways by such a division of Europe. The more that disorder develops in Western Europe, the more secure will become his control of Eastern and Central Europe, while the greater will become the prospect of his own sphere being extended westwards without effort. In sum, the 'victory' achieved automatically spells the defeat of our long-standing aim that no one power should be dominant on the Continent.

Unfortunately his criticism was founded on the erroneous premiss that the German Government had been driven to bid for peace in 1943 on the basis of evacuating all occupied territories and accepting virtually any security conditions required by the Allies. Hitler was to resign as soon as the conditions of a truce were accepted.[59]

Liddell Hart's reaction to the Allied dropping of two atomic bombs on Japan was predictable. For him it marked the culmination of Allied ruthlessness in the waging of total war. It was profoundly wrong of Britain to use it, or at least to speed its development and let the Americans use it. 'We cannot tell', he wrote on 10 August, 'whether the atomic bomb will end the war or end the world—or both—but it should at least spell the end of conscription, for it makes nonsense of that military system.' In a letter to the Archdeacon of Westminster on the 'Growth of Lawlessness in Warfare', he remarked that Hitler, for all his faults, had a greater apprehension than the British Government of the ultimate consequences to European civilization of unlimited devastation. He concluded that although Hitler could be indicted as an enemy of civilization on the political plane, his conduct of warfare was less open to condemnation in this respect than that of his opponents. This was not a pleasant admission for an Englishman but for the sake of historical honesty it had to be made. If the Archdeacon replied, his letter has not been preserved.[60]

As this chapter has demonstrated, Liddell Hart maintained a consistently critical attitude to British and Allied policy throughout the Second World War. He believed the Allies had entered the war for an unconvincing reason and that they should have been ready to negotiate peace as soon as they had demonstrated that Germany could not win the war. Even Germany's dramatic victories in 1940 did not, in his view, fundamentally change the situation, since Britain remained undefeated with the promise of American support to keep her in the war. Once Hitler's initial drive into Russia had been checked, Liddell Hart felt the case for a compromise peace was strengthened because it was now obvious that Germany was likely to lose in the long run. After mid-1943 he believed that the war had been 'won' in the only sense that mattered to the Allies, i.e. Germany's aggression had been decisively curbed, and thereafter the longer the war lasted the worse the outcome would be for Britain both strategically and economically.

From an early stage in the war even Chamberlain's Government—though the appeasing spirit still flourished behind a firm façade—followed a more bellicose policy than that favoured by Liddell Hart, for example, in its rejection of the

Russo-German 'peace proposals' in October 1939 and its cool reaction to the Dutch and Belgian monarchs' offer to mediate early in November. With Churchill as Prime Minister, once the crisis of May and June 1940 had been passed, Britain committed herself not simply to national survival but to a total war strategy designed to conquer the Axis Powers and destroy German militarism as well as Nazism. At the time many thought that Churchill's gamble would fail—Lloyd George among them. It was hard to see how David could actually kill Goliath until first the Soviet Union and then the United States entered the war. Thereafter the Grand Alliance itself required some outward sign of its members' total commitment to victory and found it in the slogan of 'Unconditional Surrender'. Indeed all that united them was the determination to defeat Hitler. The dropping of the atomic bombs and Allied occupation of all the Axis countries was the logical culmination of this policy. Liddell Hart remained unreservedly opposed to this ruthless quest for the 'mirage of victory'.

It must first be said that it required considerable moral courage to stick to his heretical views and to publicize them as widely as possible in articles, books and private correspondence. In the period of Churchill's Premiership, particularly, he was constantly swimming against the tide of popular opinion as voiced by the Press, and inevitably he received a number of abusive letters. Nor does he ever show any sign that he might have been mistaken. From an ethical standpoint his position was respectable, since he maintained that Britain's interest lay in curbing German aggression and showing Hitler he could not win. He was a pessimist but not a defeatist; he was always in favour of negotiations, but the terms must be honourable from a British viewpoint.

In practice, however, the differences from the Churchillian outlook are obvious. Churchill and those who supported him believed that it was impossible to treat with Hitler and the Nazis and that there could therefore be no question of negotiations until an evil regime had been destroyed. As a corollary, they had little if any faith in anti-Nazis allegedly in favour of moderate peace terms. This belief that the Nazis were uniquely barbarous and *not* comparable with other would-be masters of Europe was of fundamental importance. The development of a truly world-wide war from the end of 1941 onwards made a negotiated peace even less feasible for several reasons: a commitment to total victory was necessary to preserve the anti-Axis alliance, which was in reality fighting two separate wars; the Allies could not negotiate from the brink of utter catastrophe; and the extreme brutality of the war on the Eastern Front and in the Pacific provided strong evidence that the Axis forces were not interested in negotiations on any terms acceptable to an undefeated opponent.

Against this background Liddell Hart's ideas seem well-intentioned but utopian. Contrary to his belief at the time, there is no evidence that Hitler ever contemplated a peace without victory, and the Japanese position was, if possible, even less flexible. Even if she had no direct quarrel with Britain, and that was far from self-evident in 1940, Germany had to be defeated in the interests of European civilization. It was regrettable that a good deal of that civilization had to be destroyed in the process and that some countries could not be liberated, but it is hard to see that in broad terms there was any easier solution. Nor does it follow that—for Western Europe at any rate—victory was indistinguishable from defeat. As A. J. P. Taylor, no devotee of total war or uncritical admirer of Allied strategy, concludes with pardonable over-simplification:

> Future generations may discuss the Second World War as 'just another war'. Those who experienced it know that it was a war justified in its aims and successful in accomplishing them. Despite all the killing and destruction that accompanied it, the Second World War was a good war.[61]

Notes

1. See, for example, General Montgomery-Massingberd's virulent remarks in a speech to the Lincolnshire Branch of the British Legion on 4 May 1940, 11/1940/29 (all references to numbered files in this chapter are to category 11 unless otherwise stated).

2. For a foretaste of what we may expect see Maurice Cowling's scintillating study *The Impact of Hitler: British Politics and British Foreign Policy 1933–1940* (C.U.P. 1975). For the widespread persistence, after 3 September 1939, of the desire to negotiate with Germany and avoid total war see Paul Addison's

brilliant chapter in A. J. P. Taylor (ed.) *Lloyd George: Twelve Essays* (Hamish Hamilton: Atheneum 1971). In October 1939 Lloyd George received a 'huge bag' of letters overwhelmingly supporting his speeches in favour of a negotiated peace; ibid. pp. 368–369.

3. 1939/105 'Those whom the Gods wish to destroy they first make mad' 10 September. Ibid. 110b typed diary notes 2 October. Ibid. 131 13 November. Ibid. 145 n.d. 'Some foolish prophecies'. 1940/51 'King Leopold's Capitulation—a study in hysterical indency' 28 May. Ibid. 138 extracts from the *Sunday Express*, 1940–42. Ibid. 142 'Rolls of Dishonour' n.d. holograph. For Keyes' and Voigt's views see *This Expanding War* (1942) p. 265. Voigt was editor of *The Nineteenth Century*.

4. 1939/99 'The Problem if Hitler should make a Peace Offer' 7 September. Ibid. 100 'The Need for a New Technique of War' 8 September. Ibid. 109 'A Personal Problem' 14 September. Ibid. 110b Diary Notes (typed) 18, 19, 28 September.

5. *The Current of War* (February 1941): see especially pp. 151–163.

6. Ibid. pp. 145–149. 1939/108b 'A Personal Conclusion' 10 September. 1940/69 'An Historical Eye-view' 4 July. 1941/38 'The Present War in Historical Perspective' (holograph) 14 July. *Why Don't We Learn From History?* (1944) pp. 34–40.

7. Cowling op cit. p. 298. See also D. C. Watt 'The Initiation of the Negotiations Leading to the Nazi–Soviet Pact' in C. Abramsky (ed.) *Essays in Honour of E. H. Carr* (Macmillan: Shoe String Press 1974).

8. 1939/110b Diary notes 2 October. 1940/63b Diary notes 16–29 June.

9. 1939/118b Reflections 8 October. 1941/58 Diary note 4 September. Lloyd George to Liddell Hart 20 November 1939. Frances Stevenson to Liddell Hart 29 November 1939. Liddell Hart to Lloyd George 10 September 1941. See also *The Current of War* p. 163: *Memoirs* II p. 275.

10. 1941/41 a & b 'A Reflection of Freedom' (partly typed) 7 August. Ibid. 47 28 August. 1942/25 a, b, c 'Conscription kills enthusiasm' 2 April. 1943/3 'A Reflection' 30 January. *Why Don't We Learn From History?* pp. 21–30.

11. 1939/1 Diary notes (typed) 21, 28 October, 1 November. Ibid. 128b 'The Prospect in this War' 7 November. Lord (Robert) Cecil to Liddell Hart 22 November 1939. Liddell Hart's views on a compromise peace during 1939–40 were very similar to Lloyd George's, though not perhaps founded on the latter's fervent personal admiration for Hitler. See Addison op. cit. pp. 361—370 *passim*.

12. 1939/137 'The True Object in War' 21 December. 1940/6 Diary notes 20 February and 1940/9 23 February. *Memoirs* II pp. 259–260. 4/11 Ministry of Information. Correspondence with Frank Darvall 18–21 March 1940. J. Luvaas *The Education of an Army* (U. of Chicago 1964: Cassell 1965) p. 418n prints Darvall's request but not Liddell Hart's response.

13. 1940/8 Reflections 14 February. Ibid. 11 'The Unseen Checks on Aggression' 5 March.

14. Ibid. 15 'Sundry Notes' 6–8 March. Ibid. 22 'Irrationalism' 6 April. Ibid. 32 memorandum for Lloyd George 6 May. Ibid. 33 note for Lloyd George 9 May.

15. Liddell Hart to Lloyd George 20 April 1940. 1940/36 Diary note 8 May.

16. Ibid. 37 11 May; 73 Notes for History 18 July; 74 talk with Lloyd George 23 July (holograph); 79 talk with Tom Jones 31 August (holograph). As late as 11 October Liddell Hart dropped a strong hint that Lloyd George might lead a campaign for a 'Recall to Sanity'. On Churchill's eagerness to persuade Lloyd George to enter the War Cabinet and his reasons for refusing, see Addison op. cit. pp. 372–376.

17. See, for example, 1940/48 'An Historical Sidelight on the German Offensive of May 1940' n.d.; and ibid. 56, 57 and 65.

18. *Dynamic Defence* (November 1940) pp. 12–20.

19. See R. H. S. Stolfi 'Equipment for Victory in France in 1940' in *History* February 1970.

20. *The Current of War* p. 399.

21. 1940/87 'War and Peace' 14 September.

22. Ibid. 90, 91 'Simple Arithmetic (Calculations) for Statesmen' 12 October. John Brophy to Liddell Hart 19 November 1940.

23. Ibid. 102c 'The Reckoning' 17 November.

24. Ibid. 104 'Seven Pillars of Wisdom' 24 November. *Why Don't We Learn From History?* p. 53. See also Liddell Hart to Bishop Woods 20 August 1940.

25. *Dynamic Defence* pp. 53–56: cp. *The Current of War* pp. 405–406.

26. 1941/18 Reflections 17 March. Ibid. 21 'Passchendaele in the Air' 12 April.

27. *This Expanding War* pp. 70–73, 79, 101–104, 266–269. *Why Don't We Learn From History?* p. 41.

28. *This Expanding War* pp. 145, 161, 177.

29. 1941/54 'The Eternal Will O' the Wisp' 3 September.

30. Hore-Belisha to Liddell Hart 10 September 1941. J. Brophy to Liddell Hart 17 September 1941.

31. 1941/57 'The Will O' the Wisp and the Precipice' 23 September. J. Brophy to Liddell Hart 2 October 1941. Liddell Hart to J. Brophy 11 October 1941.

32. 1942/1 Diary notes (typed) February and March *passim*. Ibid. 13 'Odd Notes' 17 March. Ibid. 4 talk with Sir A. Rowlands 18 March. Ibid. 30 talk with Sir A. Rowlands 30 April. Ibid. 32 talk with P. J. Grigg 1 May.

33. Ibid. 18 talk with Maisky 18 March. Ibid. 22 talk with Lloyd George 22–23 March.

34. Liddell Hart to Bishop Woods 12 April 1942. For the gulf between Liddell Hart's thinking and the Government's see, for example, 1942/36 'The Three Stages of Military Thought' 18 May. In 1941 Richard Stokes, Labour M.P. for Ipswich, a Catholic, an arms manufacturer, an amateur diplomat and chairman of the Peace Aims Group, invited Liddell Hart to address a group of dissatisfied M.P.s at a meeting to be chaired by Sir John Wardlaw-Milne. Nothing came of it as far as Liddell Hart was concerned; Lloyd George correspondence, Richard Stokes to Liddell Hart 7 April 1941. See also Addison op. cit. pp. 370, 375–376.

35. 1942/15 talk with Beaverbrook 19 March.

36. Ibid. 37, 42, 78, 82 talks with Beaverbrook 23 May, 11 June, 23 and 26–27 September.

37. 1941/26 Notes for History 30 May; and ibid. 78 19 December. Liddell

Hart to Bishop Bell 16 October 1940. Bishop Bell to Liddell Hart 31 October 1940. Liddell Hart to Bishop Bell 2 November 1940.

38. 1942/40 'A Reflection arising from the Cologne Raids' 2 June. *This Expanding War* pp. 43, 73.

39. 1943/41 talk with Pile 4 July; 42 talk with Rowlands 7 July; 43 talk with Peck 7 July. For an illustration of Churchill's fierce resolve to '*do anything* rather than go down' see A. J. P. Taylor (ed.) *Off the Record: Political Interviews 1933–1943 W. P. Crozier* (Hutchinson 1973) p. 211. Churchill was here talking to Crozier about the attack on the French Fleet at Oran.

40. 1944/29 talk with Alexander Clifford 4 April. 1945/4 holograph note 7 April. For a concise, sympathetic account of Bomber Command's problems see Noble Frankland *The Bombing Offensive Against Germany* (Faber 1965).

41. 1943/2 Sundry Reflections 17 April, 10 June.

42. Ibid. 47 'The Background to Unconditional Surrender' 31 July; and 49 'Note on Unconditional Surrender' 21 August. Liddell Hart to Bishop Bell 12 February 1943. Bishop Bell to Liddell Hart 24 February 1943.

43. 1943/39, 40 'A Reflection on Strategy and Policy—and Humanity' 3 July. See also Sir A. Rowlands to Liddell Hart 11 July 1943 in which he suggested that modern technical skills would enable the physical basis of civilization to be restored much more quickly than would have been possible even twenty-five years ago. He did not of course imply that this constituted a moral defence or extenuation of strategic bombing.

44. 1942/59 Reflections 'Hitlerism' November 1942.

45. Ibid. 68 Notes for History 28–29 August. Bishop Bell to Liddell Hart 25 September 1942 and 4 February 1943. Liddell Hart to Bishop Bell 22 March 1943.

46. It is essential to distinguish between the noble aspirations and heroism of some of the conspirators and their efficiency. For the problems of the British response see C. Sykes 'The German Resistance in Perspective' in *Encounter* December 1968; D. Astor 'Why the Revolt against Hitler was ignored' ibid. June 1969; and comments and correspondence in the July, August, September and October issues 1969. On von Paulus' promotion see General Zeitzler's chapter p. 163 in W. Richardson and S. Freidin (eds) *The Fatal Decisions* (Joseph: Sloan 1956).

47. 1943/60–63 'The Future Balance of Europe' 1 October.

48. Ibid. 76 7 December.

49. G. Faber to Liddell Hart 16 November 1943. Lord Geddes to Liddell Hart 1 November. G. Murray to Liddell Hart 17 December. J. Brophy to Liddell Hart 8 November. Liddell Hart to J. Brophy 9 November. Oliver Harvey to Liddell Hart 25 January 1944. 1944/9 talk with Cripps 5 March.

50. 1943/75b Reflections on 'War Aims' 1 December 1943 and 'Victory' 12 January 1944.

51. 1944/17, 20, 23, 26.

52. Liddell Hart to J. Brophy 25 July and 9 August. J. Brophy to Liddell Hart 30 July and 17 August 1944.

53. I have come across no wartime references in the Liddell Hart Papers to the Katyn Forest massacre, but here was a case where the Germans were almost certainly wrongly accused of mass-murders perpetrated by the Russians.

54. 1942/59 'Codes of War' 4 September; see also ibid. 'On the Responsibility for War's Evils' 15 July.

55. See *Why Don't We Learn From History?* pp. 51–53. Yet he noted Clifford's report that American and French Colonial troops had committed atrocities in Sicily, 1944/29 op. cit. For an account of the German massacre of officers and men of the Royal Norfolk Regiment in France in 1940 see Cyril Jolly *The Vengeance of Private Pooley* (Heinemann 1956).

56. *This Expanding War* p. 185.

57. 1945/15 Talk with Sir Hartley Shawcross 28 September.

58. 1943/69 'A Fall from Reality' October.

59. 1945/5 'Sense of Proportion' 7 April. It is interesting that after the war he believed from a chance remark by a German prisoner-of-war in England that there had been a meeting between Ribbentrop and Molotov to discuss the ending of the war in 1943, and repeated this as a fact in his *History of the Second World War* (1970) p. 488. See Stephen Brooks' note on this episode at end of 1945 *Talks and Memoranda* file, unnumbered.

60. 1945/10 'Some Reflections on the Atomic Bomb' 10 August; and ibid. 11 'Reflections on the Growth of Lawlessness in Warfare' 20 August.

61. A. J. P. Taylor *The Second World War* (Hamish Hamilton: Putman 1975) p. 234.

6

The Revolution in Warfare, 1945–50

The rapid deterioration of Anglo-American relations with Russia after 1945, and Britain's equally rapid decline as a great imperial Power, seemed to bear out Liddell Hart's pessimistic predictions during the war years. As he wrote, in 1948, in a reflection on the European situation:

> We have been through a year that produced far more evil than good. It could be seen from the start—though our 'blind leaders of the blind' did not see it—that the war was bound to result in futility if it ended in victory, as desired, for the overthrow of Germany would entail the predominance of Russia in Europe, thus raising a further and greater danger to the West. At the same time the *total* character of the prolonged struggle was bound to endanger our deeper aims, since we sacrificed so much freedom in 'fighting for freedom', while the post-war circumstances were bound to make its recovery more precarious...
>
> We are now faced with the prospect of another total war against a still more totalitarian power. Such a conflict, and even the preparation for it, is likely to carry worse evils in its train, without bearing any good promise in the event of victory.[1]

Both in military circles and with the wider readership of his journalism Liddell Hart's stock began to rise, albeit slowly, after the nadir of the war years when he had deliberately cast himself in the role of Churchill's critic. Now, as after 1918, the fruits of victory seemed less sweet than in anticipation. Now also it was widely recognized that his tactical and strategical ideas had been exploited with brilliant results by some of the German generals, while some British generals also acknowledged his influence in the later stages of the North African campaign. At the end of the war, by a remarkable coincidence, many of the captured senior German generals were incarcerated near Liddell Hart's home in the Lake District and he was able to interview them at length about their conduct of operations and relationship with Hitler. These interviews formed the basis for Liddell Hart's important pioneering study *The Other Side of the Hill* (1948).

Once again, moreover, he found the political climate in Whitehall receptive to his ideas. He was in close touch with a number of Labour Members of Parliament including Richard Stokes, John Strachey, Richard Crossman and George Wigg; most important, he was consulted on several occasions by the Secretary of State for War, Emanuel Shinwell, mostly on the perennial problem of how to obtain a more efficient Army at reduced cost.

On the other hand he was bitterly disappointed when passed over for the Chichele Professorship of the History of War at Oxford in 1946. He admitted privately to Lord Hankey and Bishop Bell, who had both supported his candidature, that he had had his eye on the Chair for twenty years (i.e. since its first occupant Spenser Wilkinson vacated it in 1925), but feared he had offended too many of those making the appointment by the stand he had taken during the war.[2] In any event the Chair went to a competent military historian, Captain Cyril Falls, one of the official historians of the First World War, who had succeeded Liddell Hart on *The Times* and could be considered orthodox and 'safe' but was patently lacking in Liddell Hart's range of interests, Herculean industry and originality. In retrospect one must surely regret that at this stage of his career Liddell Hart was deprived of a secure academic base from which to write his history of the Second World War and fulfil his considerable gifts as a teacher.

In the spring of that same year, 1946, the Liddell Harts moved from the Lake District to an elegant Georgian house at Tilford in Surrey, where they spent two very happy years. When this proved too expensive for a freelance military critic, a smaller house was chosen at Wolverton in the north of Buckinghamshire, where they remained for ten years.

This chapter will examine just two of Liddell Hart's multifarious concerns in the years 1945 to 1950: one of them dominant in his articles and books, the other prominent in his private correspondence and letters to the Press. The former—which prompted the heading of this chapter—was his effort to adjust his general thinking about war to the new phenomenon of atomic weapons. Here, though now fifty years old, and widely recognized as a pundit of infantry tactics and armoured warfare of the 1920s and 1930s, Liddell Hart showed that he was still extremely versatile, adaptable and progressive in his outlook. He can indeed claim to be among the earliest pioneers of atomic-age theory for, as Michael Howard has justly observed, he was making prophecies as early as 1946 'which twenty years later were to be commonplaces of strategic thinking'.[3] The latter theme follows naturally from Liddell Hart's detestation of the Allies' unconditional surrender and strategic bombing policies: he opposed the war-crimes trials of enemy political and military leaders, believing them to be both unjust and impolitic. Between 1946 and 1952 Liddell Hart devoted a great amount of time and energy to opposing particular trials and sentences, and more generally protesting about the conditions in which former enemy generals were detained. Unlike one of his principal allies in this campaign, Lord Hankey, he did not display any marked interest in the Japanese 'war criminals', which is perhaps not surprising in view of his professional and personal connections with many of the surviving German generals such as Manstein, Rundstedt and Blumentritt. It would however be a gross misreading of Liddell Hart's character to suggest that his main motives were either personal or professional: his correspondence plainly shows that the sense of fair play, humanity towards the defeated and political moderation towards future allies were uppermost in his thoughts.

Though not published until 1946, the greater part of *The Revolution in Warfare* was written in 1944, with the reflections on 'The Atomic Bomb' added in the autumn of 1945. It is rather surprising that this is not numbered among Liddell Hart's best-known works, because it contains an admirable summary of his basic historical ideas which have been discussed in previous chapters of this book. While praising his views on the significance of atomic weapons, reviewers were generally unconvinced by his highly individualistic reading of the lessons of history. Admiral Sir Herbert Richmond, for example, though broadly sympathetic to Liddell Hart's concept of a 'British Way in Warfare' remarked in the *Fortnightly Review* that 'His opinions and assertions on these [historical] matters will find no support in the calm atmosphere of history—a source to which he continually advises his readers to go for their understanding of war'.[4] There were, nevertheless, many cogent arguments in the two historical sections on 'the technique' and 'the manner' of warfare.

In the former, Liddell Hart reached the persuasive conclusion that area bombing constituted an attack on the foundations of civilized life and heralded the bleak dawn of 'automatic warfare' in which the traditional qualities of the warrior no longer counted. The pilotless plane and the long-range rocket 'tore away the veil of illusion that had so long obscured the reality of the change in warfare—from a fight to a process of devastation ... the multiplication of the destructive force embodied in the atomic bomb drove home the lesson'.

> The advent of 'automatic warfare' should make plain the absurdity of warfare as a means of deciding nations' claims to superiority. It blows away romantic vapourings about the heroic virtues of war, utilized by aggressive and ambitious leaders to generate a military spirit among their people. They can no longer claim that war is any test of a people's fitness, or even of its national strength. Science has undermined the foundations of nationalism, at the very time when the spirit of nationalism is most rampant.[5]

As regards the 'manner' of warfare, Liddell Hart argued that limitations, restraints, and civilized conduct generally, had been progressively abandoned from the French Revolution onwards until the nadir had been reached in the total nature of the two World Wars. It is interesting to note that Liddell Hart's erstwhile hero as a brilliant exponent of the strategy of indirect approach, General Sherman, was now by implication a villain in that his draconian operations inside the Confederacy had achieved the war aim at the expense of the peace aim: they 'not only impoverished but poisoned the soil in which peace had to be replanted'.[6] The case of General Sherman neatly exemplifies the difficulty of reconciling Liddell Hart's two basic tenets: the strategy of indirect approach designed to paralyse the enemy's will to fight; and the need to limit war with a view to achieving a moderate peace. In his early works such as *Paris* he had tended to assume that the indirect approach would necessarily be most conducive to a speedy and moderate victory, but by the end of the Second World War the British bombing offensive had revealed the error of his assumption. What he never seems to have grasped entirely is that even when Britain's war effort had been strictly limited, the effect was often 'total' for her opponents, as for example in her nineteenth-century colonial campaigns.

Two other points in this section are worth mention. By 1945 Liddell Hart was aware of the Nazis' barbaric treatment of conquered populations during the Second World War. As he put it: 'This reached a depth of inhumanity unplumbed since the Wars of Religion, and has been more demoralizing through its elaborate organization. Scientific ruthlessness can be more deadly to civilization than savagery'. He attributed this bestial behaviour, however, strictly to the civil side and took a remarkably favourable view of the conduct of the *Wehrmacht*.

> On the military side, in contrast, the balance of evidence shows that the level of behaviour was better in a number of respects than that of the First World War, at any rate in the struggle between Germany and her Western opponents. Thus, the armies in general observed many of the rules contained in the old code of war, both as between themselves and in dealing with the civil population of areas they overran. Allegations of military atrocities were far fewer than in 1914–1918, and so were authenticated cases. This was a significant phenomenon, and the most hopeful one for the future that emerged from the war. It might be turned to good account.[7]

One is here reminded of A. J. P. Taylor's paradox that in the First World War German atrocities were universally believed in but were in fact rare, whereas in the Second World War atrocity stories were generally discounted but were in fact all too true. Even Liddell Hart's exclusion of the Eastern Front does not adequately meet the objection that the German Army often behaved with appalling savagery. He went on to repeat the arguments, discussed in the last chapter, that Britain had initiated the indiscriminate bombing of cities while it was the Germans who had attempted to preserve the accepted code of military conduct. The Allies' combination of an unlimited aim with an unlimited method was, by implication, more barbarous than the Continental practice.

It is surprising, to say the least, that these idiosyncratic views were not more sharply challenged. The distinguished *New Yorker* correspondent A. J. Liebling did however attempt to convince Liddell Hart that the R.A.F. had not 'started it'. He pointed out that Warsaw had been bombed on the second day of the war while the German army was still far away; he himself was an eye-witness of the bombing of Paris on 3 June 1940; and Bordeaux, crammed with refugees, was bombed on 18 June when France was obviously about to seek peace. (He could also have cited the German bombers' destruction of Guernica in 1937.) In his reply, Liddell Hart rightly stressed that he was not a Nazi sympathizer, but his only substantial rejoinder to the particular examples cited was that Hitler was conforming to his own restriction, proposed in 1935, that the fighting zone should extend for a distance of about 100 kilometres beyond the advanced ground troops. Thus the bombing of such cities as Warsaw, Rotterdam and Paris could be justified as tactical and not strategic.[8]

In his early reflections on the significance of the atomic bomb Liddell Hart wisely avoided an extreme position. On the one hand there was a natural tendency to believe that the new weapon signalled the end of warfare, or at least made all existing forces obsolete. On the other hand some conservative soldiers were arguing that it was merely a larger bomb to which an antidote could be found—everything would go on as before, including the armed services.[9] Liddell Hart, however, was convinced that the bomb *had* fundamentally changed the problem of security. He could not see any effective antidote in the short run, so that there would be a dangerous period when an aggressor might be prepared to take a gamble. 'It would seem,' he wrote soberly, 'that the rest of the lives of all peoples now living will be spent under the chilly shadow of "atomization" without warning.' Warfare as experienced in the last thirty years was not compatible with the atomic age.

> If one side possesses atomic power and the other does not, embattled resistance makes nonsense. That spells the disappearance of warfare in such cases. Resistance must be transferred into subtler channels, of non-violent or guerrilla type ...
>
> Where both sides possess atomic power, 'total warfare' makes nonsense. Total warfare implies that the aim, the effort, and the degree of violence are unlimited. Victory is pursued without regard to the consequences. In the chaotic aftermath of the Second World

> War, we are beginning to realize what the lack of any prudent limitations has meant in the way of stultifying our objects. That recklessness has left us not only impoverished but faced with harder problems than before. An unlimited war waged with atomic power would make worse than nonsense; it would be mutually suicidal.
>
> That conclusion does not necessarily mean that warfare will completely disappear. But, unless the belligerent leaders are crazy, it is likely that any future warfare will be less unrestrained and more subject to mutually agreed rules. Within such limits it may develop new forms.[10]

He went on to argue that a future aggressor aiming at profitable expansion would probably hesitate to employ atomic bombing (as counter-productive) but might try to intimidate an inferior opponent by the mere threat of using it. But more likely he would seek to achieve a very quick paralysis of the opposition: thus under the shadow of the atomic bomb there could well be an expansion of highly specialized, mobile forces.

In an extremely perceptive paragraph he suggested that future aggressors might develop the technique of indirect approach—such as practised by Hitler up to 1939—so as to avoid a direct confrontation. He envisaged a 'camouflaged war' in the diplomatic field followed by strategic operations against satellites or colonies:

> 'Infiltration' would be the basic method, extended much further than it has yet been, and employed in subtler ways. The deeper and more widespread the infiltration, the more it would tend to check the employment of atomic bombing in retort. The difficulty would be increased in the case of an infiltration which took civil forms rather than open military action. The aggressor could increase his prospects of immunity if he could devise ways of luring his prospective opponents into giving 'hostages' of some kind, or of intertwining his bases with theirs so that the entanglement became a check on their power of retaliation.[11]

Moreover he correctly foresaw that a competition in atomic devastation would be so patently suicidal for all parties that the victim of conventional attack might be unwilling to loose such a catastrophic means of turning the tide. This natural hesitation might be fatal in allowing infiltration to progress too far to be resisted.

Was there any practical solution to the danger of atomic attack? In default of international control of atomic power, he argued that peacefully minded nations should concentrate primarily on pure *defence*; rather than rely on counter-offensives as in the past, they must seek to prevent the aggressor from attaining any serious initial success. Conventional forces, then, would still have a vital role to play; armies would now take priority since they alone could deal with infiltration, but they must be armies of an entirely new pattern characterized by mobility of supply as much of troop movement. Conscription—and this became an obsession—was a waste of time in the atomic era. For its 'fire-extinguishing role' the British Army would need not only armoured divisions but also airborne divisions to replace those of the traditional infantry. These were highly prophetic insights but in some respects Liddell Hart was understandably too futuristic. Before long he wrote, for example, rocket-borne supply might replace the use of transport aircraft.[12]

In terms of domestic service politics, as distinct from theory, Liddell Hart made the radical suggestion that the Army's actual strength could be increased within the present budget by savings on the other two services. The heavy-bomber force would become superfluous in the rocket and atomic age—a very futuristic pronouncement. What the Air Force needed now were fighters for defence and fighter bombers for Army co-operation. He rightly argued that the Royal Navy should cease to build battleships, but did not specifically mention submarines in referring to the need for 'lighter craft'.

In his 'Conclusions' Liddell Hart likened the problem of security to a coin—the 'head' is the prevention but the 'tail' is the limitation of war.

> If experience has taught us anything, we should now be capable of realizing the danger of concentrating exclusively on the perfectionist policy of preventing war, while neglecting the practical necessity, if that policy fails, of limiting war—so that it does not destroy the prospects of subsequent peace.

He was realist enough to see that the idea of world federation was a very distant and dubious goal, while the disunity of the so-called United Nations was already becoming apparent. As a positive check to aggression he urged once again his scheme of qualitative disarmament—first adumbrated for the Geneva Disarmament Conference of 1932—which would deprive the would-be attacker of the essential instrument of success. He was confident that from the technical viewpoint such a scheme was feasible: the great problem was whether the United Nations would agree to accept it, together with the necessary system of technical supervision.

Failing an agreement to ban aggressive instruments of war, he thought the best hope might lie in trying to revive a code of limiting rules for warfare, based on a realistic view that wars are likely to occur again, and that the limitation of their destructiveness is in everybody's interest. To a certain extent the conduct of the Korean war could be said to have fulfilled Liddell Hart's expectations in this respect. The book ended on a moralizing note on a theme with which he was preoccupied at the time; namely that the best hope for civilization lay in a revival of manners.

> For only manners in the deeper sense—of mutual restraint for mutual security—can control the risks that outbursts of temper over political and social issues may lead to mutual destruction in the atomic age.[13]

In making his first sketches of the atomic *terra incognita*, Liddell Hart inevitably included a few non-existent 'islands' while omitting other shadowy features of the terrain, but on the whole his strategic 'map' looks astonishingly accurate in retrospect. Writing before the Cold War was officially declared, he was understandably vague here—though not in private memoranda—about the balance of forces between East and West. Russia, at this date, had not yet tested an atomic bomb, while America's stockpile was very small after she had dropped her only two existing bombs on Hiroshima and Nagasaki. Even so he foresaw that mutual deterrence was likely to function between atomic powers while not removing the danger of 'infiltration' by one of them if there was a great disparity in conventional forces. Moreover he saw clearly that atomic weapons were not by themselves likely to rule out conventional warfare at a lower level of intensity than the total wars of the recent past. Above all, since he entertained little hope of a political solution through World Federation or the United

Nations, he saw the urgency of *limiting* war should it occur both by what later came to be called 'arms control and limitation' and by the revival of civilizing restraint in its actual conduct.

None of this may sound particularly original now, but the situation was very different in the atmosphere of uncertainty and growing hostility towards the Soviet Union in the immediate post-war years. As ever, Liddell Hart saw his writings partly as counterweights to the more extreme views current at the time. Whereas in 1939 and 1940 he had warned against the delusion that Britain could defeat Germany, so now in 1946 he endeavoured to 'damp down some of the war talk which is becoming all too common'. Later in 1948 he found the situation increasingly sombre and wrote to Bishop Bell:

> A warlike mood is working up everywhere—but above all in America. From what I hear, the United States Government would have no such difficulty in carrying the country into a war with Russia as they had in the case of the previous wars. While a firm line is essential in dealing with Russia, I feel it is vital to avoid an 'air of provocativeness'. We should press on with the project of Western Union while trying to avoid its becoming too definitely a satellite of America.[14]

Liddell Hart's *Defence of the West* (1950) was composed largely of articles already published between 1946 and 1949 in such newspapers and magazines as the *Sunday Pictorial*, *John Bull*, *Picture Post* and *World Review*. Though unsatisfactory as a book—it contains contradictions and repetitions and lacks a unifying theme—it nevertheless provides a useful guide to Liddell Hart's thinking in the late 1940s. As the *Manchester Guardian* justly remarked in an editorial leader, the book lacked 'pungency and shape'. 'Yet scattered among its pages, like rare plants in some queer herbaceous border, are ideas and criticisms which will stimulate all but our least intelligent soldiers.' The editor also pin-pointed two of Liddell Hart's irritating traits as an author. First, he treated his opponents with too little respect: 'He spoils a good case against conscription, for example by putting the arguments in its favour in far too feeble a form.' Second, in dealing with the lessons of the 1939–45 war, 'he has a tiresome habit of saying "I told you so" or something like it. . . . Yet this new commentary can surely do nothing but good. . . . In almost every chapter Captain Liddell Hart comes up with some sensible comment.'[15]

In posing the question 'What would another war be like?' Liddell Hart saw the weakness of the West's ground forces in Europe as the chief danger spot. There was little to stop a Russian tidal wave surging forward to the Mediterranean or the Atlantic coast, or both. Indeed, supply difficulties might check the Red Army's momentum more than military resistance. In private memoranda he displayed much interest in gases or radioactive dust as possible alternative means of defence. 'The prospect of atomic bombs dropping on Moscow and Leningrad might be a deterrent to a decision for war, but would hardly be decisive of the issue once war broke out.' He doubted whether atomic bombs could check the Red Army's progress. At present they were not a tactical weapon suitable against land forces—particularly forces so fluid as the Russians. He envisaged a long war degenerating into a stalemate, in which the peoples of Western Europe would suffer far more than in 1940. In discussing the problems of the range and accuracy of guided missiles he anticipated the employment of Polaris submarines in remarking:

> Thus seapower might find fresh offensive scope through the development of new types of vessel designed as floating platforms for guided missiles which could be launched to their target from points close to the enemy's coast.[16]

He feared that if one side had an overwhelming advantage, weapons of mass destruction would be used. If on the other hand both sides had the same weapons—or even imagined that they had—then both might hesitate to unloose them from mutual fear of retaliation, as had been the case with the non-use of poison gases in World War Two.

In discussing the question 'Could we survive another war?' Liddell Hart expressed alarm at the advice he believed military advisers were giving to the effect that another war would be of the same style as the last except for some fresh trimmings. 'The extent of apparent unconcern with the prospect suggests an amazing lack of vision and understanding.' The two great rivals, Russia and America, would suffer less but for Britain and Western Europe the outcome could only be disastrous. He was also uneasy at the way in which the Labour Government, following Churchill's wartime lead, had tied Britain's foreign policy so closely to America's. He feared in particular that George Marshall, formerly Chief of Staff and now Secretary of State, was too offence-minded. If Britain and Western Europe were to become merely an advanced base for the mounting of America's atomic and rocket weapons, their lot would be disastrous whatever the ultimate outcome. 'With ruthless candour, American defence memoranda have described Britain as America's shock-absorber in another war.' This was a fatal role. Britain should concentrate entirely on the two facets of defence—prevention and protection. For the former she required a small professional army organized in armoured and airborne divisions to check infiltration by Russian spearheads, and for the latter a vast civil defence programme designed to put essential services and industries underground in event of war.[17]

Liddell Hart's hobby-horse, the need for comparatively small (20 divisions) élite ground forces to deter a Russian incursion into Western Europe, was developed further in the chapter on 'The Defence of Western Europe', where he rightly stressed that 'it is the forces on the scene which count, not a prospective reinforcement from overseas, however great this might ultimately become'. Here he strongly challenged the opinion, recently expressed by Lords Templewood and Trenchard in a House of Lords debate, that superior air power could effectively hold up and disorganize any Russian advance. He thought excessive reliance on air power would be countered by Russia's own considerable air force, while it would be impossible to check the Red Army by this means alone, since not only was it skilled in moving dispersed but its troops were less dependent on what other armies regarded as necessities in food and equipment.

To some extent Liddell Hart had modified his views on the relative importance of air power following a lengthy exchange of letters with Air Marshal Sir John Slessor, then Commandant of the Imperial Defence College, in 1948. Slessor believed that those, like Templewood and Trenchard, who placed the main emphasis on air power were essentially correct:

I'm sure it is no good Western Union trying to build up Continental

> Armies on a scale to stop the RUSSIANS [*sic*]—why take him on at his own game even if we had the men which we have not. If we try to be strong everywhere we shall be strong enough nowhere.
>
> I am sure we must have a reasonable Army and Tactical Air Force on the ground, but am also sure that our best bet is in the air. And if I had a force on the lines Trenchard recommends I should be very happy to take on the job of stopping the horde.

He was strengthened in this view by his opinion that Russian air-maintenance capacity was very low. Liddell Hart replied that he largely agreed with Slessor's argument: it was really a matter of striking a balance between air and land forces in Europe.[18] Here is a good example of the popular journalist needing to take a more dramatic line in print than in private correspondence.

Sir John Slessor also put up powerful arguments against another of Liddell Hart's favourite proposals of that time; namely that in view of Britain's economic difficulties she should leave the provision of strategic bombing forces entirely to the United States, leaving herself free to concentrate on acquiring a strong fighter force. Liddell Hart found a supporter for his view in Air Marshal Sir George Pirie, then Air Member for Supply and Organization. Pirie pointed out that our present bomber force was useless for European defence. Its 20 squadrons were all composed of Lancasters and Lincolns and we could only afford to replace about one Lincoln per year. Slessor argued that such a complete differentiation of the roles would not be acceptable to an ally, and cited the example of France leaving strategic bombing entirely to Britain in 1939. Moreover if we had no bombing force we should have no say in the direction of a bombing campaign. Finally, it would mean in practice that we should give up research and development in the whole field and that would be unacceptable either politically or strategically. In sum, he wrote, Liddell Hart's theory was superficially attractive but in reality dangerously unsound.[19]

The chapter on '"Home Defence" in a New War' began with a sober analysis of the effects of the atomic bombs dropped on Japan and concluded that in its present state of preparedness Britain had little chance of surviving even a few atomic bombs. The unexpected rapidity with which Russia had tested her first atomic bomb—announced by President Truman on 23

September 1949—entirely upset the West's defence programme. Until recently it had been assumed that Russia would *not* develop an atomic bomb until 1952 and in the meantime land and air defences of Western Europe could be build up under the protection of the American nuclear monopoly.

> Now that the monopoly has been broken, the deterrent has diminished in value while the protection has disappeared. The peoples of Western Europe can no longer find comfort in the thought that any advance of Russia's armies could be answered by the destruction of Russia's cities. Their own cities are now exposed to similar destruction by the same means.[20]

Liddell Hart was alarmed at threatening statements of what in the early 1950s became the official American policy of 'massive retaliation'. He recorded Lord Trenchard, for example, as declaring that 'We must be prepared to say now that if this threat from the East materialises we will at once hit with the latest development in the possession of the Western Powers—the atomic bomb'. Liddell Hart rightly pointed to the enormous casualties which Russia had sustained in the recent war without surrendering. 'To start using atom bombs on our side would be the most likely way to ensure that history's verdict on our vanished civilization would be "suicide while temporarily insane".' He countered these offensive threats with an appeal for a massive civil defence programme which should receive priority even over housing. To give some idea of its scope:

> Under the great cities, a series of air-conditioned honeycomb towns might radiate from the underground railways, which could serve their present purpose of intercommunication. Provincial centres would also be provided with a low-level layer where work and life could carry on. Surface buildings, including houses, might be connected by underground passages to the central subways. Pipelines might be developed more widely, and in new ways, for the transit of supplies.

When he first submitted this article to the *Sunday Pictorial* the editor was alarmed by its pessimistic tone and asked the author if he could not find some grounds for hope. Liddell Hart noted in his diary that half-way through writing it the situation seemed to him almost hopeless: 'that the only sensible advice was to tell people either to "put their heads in a gas oven" or to emigrate'—but in the end he felt the chances of averting atomic destruction were better, provided proper measures were taken.[21] It was perhaps naïve of him to hope that such a programme would be launched, not only in view of the country's economic difficulties but also because of the alarm it would generate.

One of the most iconoclastic views advanced in *Defence of the West* was that an atomic war would not necessarily be short; indeed a newspaper heading of a similar article ran 'Despite Atomic Bombs and other Modern Weapons Another War would probably be more prolonged than the last'.[22] Liddell Hart pointed to repeated historical delusions that the next war was bound to be short. Atomic bombs would speed up the pace of destruction and make it more horrifying but they would not ensure a swift end to the struggle. Even if a number of cities were destroyed, that would not necessarily bring about surrender. Human beings have an infinite capacity for adjusting themselves to the gradual degradation of their living conditions. The lower a people's standard of living, the easier it would be to maintain its endurance. Thus, in the nuclear age, large, sparsely populated countries with a low standard of civilization would be at an advantage compared with, say, the densely populated and industrialized countries of Western Europe. Fortunately, this idea has not yet been tested in nuclear war, but the successful resistance of North Vietnam illustrates its relevance in an intensive conventional war.

The same chapter on 'The Shadow of War with Russia' contained a brilliant insight which anticipated the limitations of the Dulles doctrine of 'massive retaliation'; namely that the possession of nuclear power did not necessarily confer effective diplomatic leverage *vis-à-vis* non-nuclear countries:

> In theory, America should be able to decide every international argument as she wishes. In practice, she has not been able to get her own way—because those who oppose her policy shrewdly realise that she could only use the atomic bomb as a last resort. Opposing countries can thus feel reasonably safe in defiance over any political issue that is not a life-or-death matter to the atom-owner. They may be able to go a long way in pursuing their own aims, even aggressive aims, without deferring to her protests.

This accurate prediction was developed further in a draft article written as early as 30 April 1947 entitled 'Will the Atomic Age be Tyrant or Leveller?' where Liddell Hart wrote:

> The limitless destructiveness of such a weapon forms its own practical limitation. Its employment cannot be regulated or adjusted to circumstances like that of troops. It can only destroy; it cannot infiltrate or occupy. This condition handicaps its possibilities of being used to exert a moderate, and carefully controlled, pressure in the field of power politics. Other states are therefore fairly safe in opposing the possessor of the atomic bomb in any matter that stops short of being absolutely vital to her. For even the most steely of statesmen might well feel uneasy about the consequences of releasing such a 'mass killer', especially in a period of nominal peace or 'phoney' war. Taking advantage of that natural hesitation, other states are able to disregard protests and go a long way in pursuing their own contrary purposes.[23]

Liddell Hart's coolness and moderation were seldom displayed more impressively than at the conclusion of his essay on 'The Shadow of War with Russia' where he warned against those hotheads, more common in triumphant America than war-weary Britain, who, believing that war with Russia was bound to come, were inclined to force the issue. This school of thought

> talks of the importance of being ready 'to strike first', regardless of the basic fact that American comparative remoteness entails delay in exerting her weight, and of the risk that war might be needlessly precipitated in the attempt. It ignores the likelihood of initial Russian successes, the long road of recovery in consequence, and the irreparable damage that civilisation would suffer in the process. It underrates the difficulties of gaining so-called 'victory' over Russia even when the balance of strength has turned.

It was all too easy to precipitate war in a mood of exasperation. The chief danger lay in producing a situation where the other side could not draw back without loss of face. Despite Russia's abrasive statements and provocative actions we should not discount the underlying desire for peace among a people who had suffered so much from war.[24]

The extracts cited above are designed to convey a general

idea of Liddell Hart's broad thinking in 1950 about the problems of war in the atomic age. Such selective treatment is evidently unsatisfactory in some respects and it is a pity that Liddell Hart did not winnow, revise and rewrite his articles in compiling *Defence of the West.* As *The Listener*'s reviewer aptly remarked: 'It is the frequent sentences or paragraphs which arrest the attention and stimulate thought that give the book its interest and value rather than the design and argument of the whole. This is a matter for some disappointment because Captain Liddell Hart often sees far into the fog of war....' There was in particular in this volume an uneasy juxtaposition between Liddell Hart in his various roles as military historian, theorist of war and contemporary commentator. It is also far from clear what kind of audience the author had in mind; most of the contents had after all first appeared in rather light-weight popular magazines, nor was there enough technical information to impress the professional service experts. The reviewer in *Public Opinion* concisely characterized the book and its author when he wrote 'It is full of disturbing ideas, some shrewd and some slightly naive ... The whole is a curious hotch-potch. One thing at least we know from experience. In his way the man has a touch of genius.'[25] Further discussion of Liddell Hart's assets and limitations as a theorist of nuclear war may conveniently be deferred to the next chapter.

Between the end of the Second World War and the early 1950s Liddell Hart probably devoted more time and effort to opposing war-crimes trials and questioning the treatment of imprisoned German generals than to any other of his numerous concerns. Such a strong commitment was entirely consistent with his character and philosophy. He was a passionate believer in moderation and reason in public as well as in private life, and he held that a defeated enemy should never be victimized or humiliated. As we have seen, he believed that the German Army had on the whole behaved well—perhaps even better than the Allies—and his conviction that by and large the outstanding German generals were dignified, honourable men was strengthened by meeting some of them as prisoners of war. Lastly, as has been stressed before, he was by temperament inclined to put the opposition case to the prevalent opinion of the day on any topic which engaged his interest. In this particular cause his chief supporters, and in some instances collaborators, were Lord Hankey, Bishop Bell, Reginald Paget (Queen's Counsel and Labour M.P.), Montgomery Belgion (lawyer and author) and Professor Gilbert Murray.

Liddell Hart's attitude to the German generals was made clear in correspondence in 1946 with a Jewish friend, Vivian Gaster, who not surprisingly did not believe in leniency towards the enemy. Gaster suggested that Liddell Hart was viewing them too benevolently and too much from a professional angle. They knew very well what was happening to those whom they left to the tender mercies of the S.S. and should bear their full responsibility. He added (correctly, in my opinion) that most of the generals who had belatedly conspired against the regime were prompted by the Nazis' mishandling of the war rather than their crimes against humanity.

Liddell Hart replied: 'My attitude to the German generals is not one of benevolence, but one—I hope—of understanding: understanding human nature and the situation in which they were placed.' Between the wars, he argued, the German generals 'almost fell over backwards' in their efforts to keep out of politics. During the war he doubted if generals of any other nation would have behaved any better in their position. Indeed he was surprised at the degree to which they *did* protest (citing for an example General Blaskowitz's brave though futile protests at S.S. activities in Poland), and wondered if Allied soldiers would have done as much.[26]

To a German friend who wrote in 1949 complaining about British treatment of his countrymen in their occupied zone, Liddell Hart replied moderately that he personally was sympathetic but one must remember the attitude of the British people:

> Both wars were fought in a bitter way, and the end of the second one was marked by the revelations about the concentration camps in Germany, which created a tremendous wave of indignation here. That came on top of all the other actions which had aroused deep anger. For a time it swamped all arguments for a moderate peace . . .

The Americans were more inclined towards a harsh peace but British anger was subsiding quickly. He was well aware of a German tendency towards self-pity and remarked that Manstein was one of the few Germans he had met who displayed not a trace of it.[27]

Liddell Hart's Papers contain numerous references to support his belief that the German Army's conduct had generally been good whereas the Allies' had often been disgraceful. Alexander Clifford told Liddell Hart, for example, that an Australian had boasted to him and a fellow-journalist, Edward Ward, that he and his comrades had roasted a Stuka pilot in Crete. Clifford believed that the Australians did a lot of this sort of thing: 'They were a brutal lot of men and they made up their rules as they went along.' Cyril Ray, the *Manchester Guardian*'s military correspondent in Italy, had been appalled by Allied brutality, particularly the Canadians'. Liddell Hart also heard that Patton's troops, particularly the 45th Division, had committed atrocities in Sicily. By contrast he noted that General Martel, who had questioned inhabitants of the liberated areas of Russia in 1942, had formed the impression that the Germans had behaved well so long as there had been no trouble, but had taken drastic reprisals when their soldiers had been shot by armed civilians. The Russians with Martel did not like to hear anything favourable said of the Germans.[28] Reginald Paget took a more cynical—or realistic—view about the German generals' innocence. 'Frankly,' he wrote to Liddell Hart, 'in a total war I think that there are very few generals who would really be in a position to rebut a charge that they had been responsible for authorising killings that could not be justified by the rules of war.'[29]

Liddell Hart listed eight points in his 'Case against the War-Trials':

1. One-sided trial of vanquished by the victors—boils down to convicting them because they lost the war!
2. Hypocrisy of sitting in judgement along with the Russians ...
3. Asininity of convicting the Germans of what was done in Russia —without regard to Russian methods of war (and Katyn—the earliest case).
4. Disregards evidence of the occupied countries that the German was the best behaved army.
5. Disregards the way that Hitler and the Nazis (see Goebbels' diary) continually complained of the opposition of the German generals and of their being too humanitarian.

6. The worst German atrocity was the mass killings (by Nazis)—but how is this worse than our own mass bombing policy, from Jan. 1942 deliberately aimed at civil population.
7. As to condemnation of reprisals—compare the far more drastic American action of calling for the bombers to blot-out any village from which a shot was fired.
8. Hypocrisy of holding the Commando order against the Germans [i.e. that they should be shot] in view of the way the Commandos broke rules of war and killed defenceless prisoners.[30]

Liddell Hart's press campaigns against war crimes and the treatment of prisoners of war fell into five phases. The first began with a letter to *The Times* published on 18 October 1946 entitled 'Judgement at Nuremberg' in which, after discussing the dangerous precedents set by the victors trying the vanquished, Liddell Hart focused attention on the disparity between some of the charges and the sentences. Albert Speer's sentence of twenty years' imprisonment seemed to him unduly severe, but his chief criticism was directed at the death sentence imposed on Alfred Jodl, whose role, in Liddell Hart's view, had been merely that of a 'super clerk' signing Hitler's orders. *The Times* published dissenting letters from two eminent lawyers, A. L. Goodhart and G. D. Roberts, Recorder of Bristol and a prosecuting counsel at Nuremberg. But Liddell Hart received several private letters of support including one from the well-known authoress Vera Brittain, who remarked: 'Nuremberg was, of course, nothing but elaborately camouflaged vengeance which sets a terrible precedent. It was particularly monstrous that Jodl should have been executed, and, to a smaller degree, Keitel too.'[31]

The second phase of Liddell Hart's newspaper protests concerned the conditions in which some senior German generals, not yet indicted for war crimes, were being detained in Britain. Liddell Hart had been granted access to the German prisoners of war in 1945 but when they were removed to South Wales early in 1946 objections were made to his visits by 'hun-haters'. One grievance was that he had lent a mattress to Rundstedt, who was ill.[32] A letter to *The Times* in May 1946 describing the degrading conditions in which six full generals were confined in a small room etc. was unpublished pending the outcome of the Nuremberg trials; but in August 1948 he found a far more sympathetic editor in A. P. Wadsworth of the

Manchester Guardian. On 21 August that liberal newspaper published Liddell Hart's eloquent plea for better living conditions and more humane treatment of certain German officers, Field Marshal von Manstein in particular, who had recently been taken back to Germany as witnesses. In a strong supporting leader the editor remarked:

> No one can read the letter from Field Marshal von Manstein, quoted by Captain Liddell Hart..., without a feeling of embarrassment. Whatever one may think of German generals as a class, whatever may be said against these particular generals as individuals, their treatment is repugnant to our sense of justice. After three years in this country as prisoners of war, these three old men [Brauchitsch, Rundstedt and Manstein] have been sent back to Germany at last—only to be kept in close confinement, constantly supervised, and only allowed to see their wives in the presence of an officer. Why? They have not been tried. If they are war criminals they should have been tried long ago or not at all ...

Probably as a result of this publicity, the field marshals' prison conditions were greatly improved, but in reporting this on 10 September Liddell Hart now sought to gain support for opposition to their probable trial as war criminals. He urged that wider issues be taken into account, such as the Allies' terror bombing and the dropping of the atomic bombs. The three field marshals were old and ill and, even more important, the evidence against them was flimsy. In the *Manchester Guardian* Liddell Hart's numerous supporters included Gilbert Murray, T. S. Eliot, Osbert Sitwell and Lord Tweedsmuir,[33] while a similar campaign in *The Times* was endorsed by Gilbert Murray, Lord Parmoor, Victor Gollancz and a group of Socialists including Michael Foot, J. B. Priestley, Lord Russell, Richard Stokes and Reginald Paget. Liddell Hart attributed the ineffectiveness of this impressive correspondence, and of protests in both Houses of Parliament, to their being overshadowed by the General Election.[34]

Liddell Hart was particularly interested in the fate of von Manstein. He rated him the ablest German general of the war, and to this professional esteem was added personal admiration after he had interviewed the Field Marshal in captivity after the war. Thereafter they corresponded, though the Field Marshal was restricted to one letter per month, his family included. When he was put on trial in 1949, Liddell Hart kept in close touch with the case through his friend Reginald Paget, who had volunteered to serve as Manstein's senior defence counsel. On 28 November 1949 Paget reported to Liddell Hart that 'the Manstein trial has gone most awfully well, and I am very hopeful of an acquittal'. He had been most favourably impressed both by Manstein's dignified bearing throughout the trial and also by his quickness of mind. Liddell Hart, Paget and Hankey were consequently shocked when Manstein was sentenced to eighteen years' imprisonment, essentially for responsibility for orders by Hitler and the High Command executed by his subordinates. Hankey thought, however, that public opinion was swinging against the policy and that there was 'much real uneasiness'.

According to Liddell Hart, his letter in *The Times* on 11 January 1950 was the longest the paper had ever printed. In it he pointed out that Manstein had been found guilty on only two out of the seventeen original charges and those were not major crimes. In fact Goebbels had warned Hitler that Manstein and Kleist were disobeying his orders and the former had taken the initiative in trying to mitigate inhumane orders. 'His condemnation,' the letter concluded, 'appears a glaring example either of gross ignorance or gross hypocrisy.' Lord Hankey was perhaps more accurate in suggesting that there was no villain:

> The truth is that the whole party, the President, the members, the Prosecutors, the Defending Counsel and above all the unfortunate prisoner were victims of *ex post facto* laws and procedure, which made a defence almost impossible.[35]

The general public seemed to share Liddell Hart's view that an injustice had been done in Manstein's case, for he received about a dozen letters of support to only one of abuse. In 1952 Paget informed Liddell Hart confidentially that Manstein and Kesselring were to be released to have operations and then allowed out indefinitely on convalescence. Apart from the difficulty of dealing with the Russians over the war criminals in Spandau (where the pitiless Russian attitude has kept Rudolf Hess incarcerated to this day), Paget thought that the chief stumbling-block to the release of German generals was Anthony Eden—not because he was inhumane but rather that he regarded the United Nations idealistically and viewed the war-crimes trials as part of a noble experiment to establish universal law.[36]

After a period when the immediate post-war hatred of the Germans seemed to be abating, public hostility was again encouraged by certain newspapers in 1951, possibly because of apprehension about plans to re-create a West German army as a vital element in European defence. Liddell Hart wrote to Paget that he was at a loss to understand the 'anti-German note now widespread in our papers' and the outcry against the reductions of sentences of war criminals proposed by the Americans. 'Moreover,' he added characteristically, 'no one seems to be raising their voices on the other side, in favour of justice and mercy.'

Never lacking in courage in such cases, Liddell Hart published a long letter in the *Manchester Guardian* on 12 March 1951 which concluded:

> From the tone of your recent leaders, and many other comments like them, it would appear that we as a nation care little about the consequences of continuing to trample on the German people's feelings. But we might at least show care for England's reputation of being just and humane.

The editor commented that because there had been errors or excesses in some cases (like Manstein's or the Belgians' trial of General Falkenhausen), it did not follow that we should oppose all trials:

> On the general issue Captain Liddell Hart ignores what seems to us the main point—that to bow to the German clamour is to play into the hands of the worst elements in revived German nationalism and to rehabilitate the militaristic spirit in Germany.

Liddell Hart replied that he was not opposed to war-crimes trials *per se*: if the Allies had shown care in selecting clear cases and avoiding bad grounds in pursuing the trials we should have less causes to feel ashamed. 'It is essential that we should clearly show that we are striving to be just, not vindictive.'[37]

By 1951 a new consideration was entering into Liddell Hart's concern at the treatment of ex-enemy generals: namely the controversy over whether a West German army should be created as part of a European Defence Union. As we have seen, even during the Second World War Liddell Hart had appreciated the vital need for a German army if there was to be any effective defence against the westward expansion of Russia, and he never wavered in this conviction despite the contrary view of such an important general as de Lattre de Tassigny.[38]

Consequently when a *Times* leader on 22 September 1951 stated that, except in North Africa, 'the German Army earned a terrible reputation', Liddell Hart was provoked to write a forceful rebuttal of what he believed to be a false charge:

> In visiting the countries of Western Europe, and asking people about their experiences under the German occupation, one is told over and over again that the Nazi regime was hateful but that the German Army was 'most correct' in behaviour. Indeed it is disturbing to find how commonly the personal conduct of many members of the armies of liberation is compared unfavourably with that of German soldiers, by those who had direct experience.

He then broadened the scope of his counter-attack by questioning the nature of the trials and suggested that few of the charges against German generals would have been upheld by an impartial court. The present opposition to the generals' early release was not the way to go about making the Germans 'good Europeans'. This letter, not surprisingly, in turn provoked some critical private correspondence, some writers accusing him of pro-German bias while others recounted particular atrocities perpetrated by German soldiers. Liddell Hart's courteous and moderate replies (except to abusive letters, which he ignored) show him in a most favourable light. Among those who wrote strongly supporting his viewpoint was Professor Gilbert Murray.[39]

Liddell Hart continued to press the view that vengeance must stop and the Germans be treated equally if they were to play a full part in Western defence. He was fortified in these opinions by an interview with the West German Chancellor, Dr Adenauer, in June 1952. According to Liddell Hart's notes on the meeting, Dr Adenauer—who of course had no reason to love either the Nazis or the German generals—favoured an early act of clemency by Britain and the United States. Keeping the German generals in prison was in his view harming the prospects of a European Treaty.[40]

Liddell Hart's efforts on behalf of German generals and the German Army have been quoted at length not only because they took up so much of his time in these years, but also because they exemplify his willingness to espouse minority causes in which there was no prospect of financial or other material reward, and indeed from which his own reputation and sales might well suffer. Whether or not one fully shares his belief in the honourable conduct of the German Army, it is surely possible to admire his moderation and humanity in presenting the case for decency and justice towards a beaten enemy. (Incidentally it is noteworthy that although he was opposed to war-crimes trials in principle, he only publicly took issue with cases where there was good reason to doubt whether justice was being done.) To those who criticized him for taking up the German generals' cause while neglecting other worthy causes (such as the plight of their victims), he might have replied that he was expertly acquainted with the military context in which the generals had had to operate; he subsequently got to know many of them personally; and they were not in a position to speak up at all adequately for themselves. Other people could speak for causes on which they were equally well-informed. (Those who are traduced for championing the cause of animals instead of maltreated children could make a reply along similar lines.) In sum, while there is clearly room for differences of opinion on the issue of war-crimes trials and the extent of the German generals' guilt, Liddell Hart emerges with great credit for his part in this whole unhappy affair. In particular, the press campaigns described above should conclusively rebut the unjustified slur that he 'used' the German generals 'to make his way again' after the Second World War.[41] Certainly their information was invaluable to him as a historian and equally clearly he relished their tributes to his foresight and influence. But in my judgement he was too big a man deliberately to 'use' the German generals—or anyone else—to further his career.

Although this chapter has concentrated on two of Liddell Hart's chief interests in the immediate post-war years—the adaptation of his theory of war to the age of atomic weapons and the treatment of the German generals—it must be stressed that he kept up all his old professional interests and added a mammoth new one which would gradually become predominant; namely the collection of evidence for a comprehensive military history of the Second World War. He also maintained regular contacts with Labour politicians interested in defence matters. As late as 15 May 1950, for example, he recorded his impressions of a one-and-a-half-hour meeting with Shinwell in which he urged the Minister to abolish conscription and discussed the possibility of allotting more money to the Royal Air Force at the expense of the other Services.[42] In addition to recovering some of his former political influence, he also had the mixed satisfaction of knowing that the value of his theories was now being openly acknowledged abroad, particularly in Germany and later Israel.[43]

Nevertheless, on the evidence of his Papers, at the beginning of the 1950s he went through a very despondent phase. To his failure to get the Chichele Chair at Oxford was added another keen disappointment. In 1951 Aidan Crawley and Lord Pakenham made enquiries on his behalf about a peerage and he was informed that Attlee's response held out some hope, but the prospect vanished when the Labour Party was defeated in the autumn election.[44]

While not embittered by these and other disappointments, he evidently felt keenly that his abilities had not been sufficiently recognized by British Governments, military institutions or universities.

> In sum [ran a typical Reflection] over the last 30 years, my own country has only made use of a small fraction—perhaps 20 per cent or less—of my capacity for service to it. It has also left its opponents abroad to get the most profit out of my published products.

He reacted sharply when Major-General Eric Dorman-Smith, whose own prospects in the Army had been eclipsed in 1942, suggested that his career had been a long success story, by listing all the 'might have beens' between 1930 and 1951. He claimed, however, that he was not embittered because he enjoyed the pleasures of his personal freedom.[45]

In 1952 a telephone call from the American Embassy in

London inviting him to lecture at several of their War Colleges prompted the following 'Reflection on Losing Interest in a Subject':

> In the last year I have had similar invitations from more than half a dozen countries. The prospect does not have much attraction and I have been wondering why. I feel that it is due to a basic loss of interest in military things, and even in military thought. And that, I suppose, is a natural sequel to manifest fulfilment as well as to thirty-five years' intensive study in this field. I ought to have cut clean out of it after 1945 and entered an entirely fresh field, as I had the urge to do.
>
> I cannot expect to produce a revolution in warfare twice in my lifetime, and anything less that I could achieve in this way would be an anticlimax. Moreover, I have no desire to produce a second revolution, since the first was turned to profit, not by this country or its allies, but by its opponents—and changed the course of the world for the worse. That might happen again in the case of any more revolutionary ideas favourable to the offensive, and thus to aggression.
>
> There is still a chance, too, that the ideas I later developed to frustrate aggression and sterilise war may be consolidated, on all sides. The Korean war has shown hopeful symptoms.[46]

Finally, on reaching his sixtieth birthday on 31 October 1955, he penned a rather dejected Reflection to the effect that as a historian he could not get at the facts of the Second World War, while neither the British Government nor the Services seemed to require his expertise. Reginald Paget, to whom he sent a copy, wrote to reassure him that he was being unduly pessimistic. No man, said Paget, can hope to be a prophet in his own country until he is over sixty: 'My impression is that England is at last coming round to your way of thinking, and that we shall get a reconstructed Army.'[47] Paget was to prove right and in a broader sense than just Army reform, for in the next decade Liddell Hart was to receive much of the recognition and some of the honours which he had surely earned.

Notes

1. 11/1948/6 'The European Situation—Deeper Currents' 7 March.
2. Liddell Hart to Bishop Bell 17 March and 3 April 1946. Liddell Hart to Hankey 26 June 1946.
3. Institute for Strategic Studies *Adelphi Paper* no. 54 February 1969 'Problems of Modern Strategy' Part One p. 20.
4. *Fortnightly Review* July 1946.
5. *The Revolution in Warfare* pp. 36–37 of American ed. (1947).
6. Ibid. pp. 72–74, 79.
7. Ibid. p. 84.
8. Ibid. pp. 90–95. For Liebling's criticism see 10/1946/43, Liddell Hart's 'War, Limited' in *Harper's Magazine* March 1946 and attached correspondence from May and July issues. Slessor also informed Liddell Hart 31 October 1947 that his references to Air Staff doctrine in *The Revolution in Warfare* were wide of the mark.
9. 10/1945/17 'The Atomic Bomb and Britain's Post War Defence' in *World Review* December 1945 pp. 38–41.
10. *The Revolution in Warfare* pp. 99.
11. Ibid. pp. 101–102.
12. Ibid. pp. 102–105.
13. Ibid. pp. 105–114, 119. See also 11/1948/12 23 May (holograph note) 'Humanity cannot afford both free thought and free manners. Free thought with formal manners makes for steady progress. But if accompanied by careless manners it wrecks civilisation.'
14. Liddell Hart to Bishop Bell 13 November 1946 and 24 February 1948.
15. *Manchester Guardian* 29 May 1950. The editor was countering a purely destructive review of Liddell Hart's book by Sir James Grigg.
16. *Defence of the West* p. 88. For instances of Liddell Hart's interest in new gases see 11/1949/30a and b 'Note for History' n.d. but pencilled in Liddell Hart's Hand '?1948 (or 49–50)'; and in radioactive dust 11/1949/2 'Discussion with G. F. Hudson' 11 January.
17. *Defence of the West* pp. 90–98. Liddell Hart could be excused for not appreciating at this early stage of deterrence theory that it might be necessary to have large conventional forces precisely in order to keep war limited: see Michael Howard 'Problems of Modern Strategy' op. cit. p. 20.
18. *Defence of the West* pp. 113–126. Slessor to Liddell Hart 1 December and Liddell Hart to Slessor 3 December 1948.
19. *Defence of the West* pp. 202–203. 11/1949/10 'Talk with Air Marshal Sir George Pirie' 15 March. Liddell Hart to Slessor 8 January and Slessor to Liddell Hart 22 January 1948.
20. *Defence of the West* p. 130.
21. Ibid. pp. 97, 136–140, 11/1949/22 Diary note 29 November.
22. *Defence of the West* pp. 144–146; cp. pp. 376–378. 10/1947/23 'Another War Would Probably be More Prolonged than the Last' in *The Outspan* 21 March 1947.

23. *Defence of the West* p. 145. 10/1947/10 'Will the Atomic Age be Tyrant or Leveller?' 30 April.

24. *Defence of the West* pp. 149–150.

25. *The Listener* 3 April 1950. R. G. Jessel in *Public Opinion* 9 June 1950.

26. 9/24/155 (and Gaster correspondence file) V. Gaster to Liddell Hart 28 July and Liddell Hart to V. Gaster 30 July 1946.

27. 11/1949/3 'Extracts from a letter to a German friend' 28 January.

28. 9/24/155 Liddell Hart to Alexander Clifford 10 May 1946. 'Notes for History' A. Clifford 4 April 1944. 'Talk with E. J. B. Rose' (Manager of Reuters' London Office) 8 February 1949 (copy in 11/1949/4). 11/1946/8 'War Crimes—Australian Troops' 23 July. 11/1947/17 'Talk with General Martel' 22 August.

29. R. T. Paget to Liddell Hart 5 November 1948.

30. 9/24/155 'Case against the War-Trials' holograph note, n.d. but *c.* 1948.

31. 9/24/157 Liddell Hart's letters to *The Times* 18 and 25 October 1946. Vera Brittain to Liddell Hart 18 October. Liddell Hart to G. D. Roberts 24 October.

32. Liddell Hart to Bishop Bell 17 March 1946.

33. 9/24/158 *passim.*

34. Ibid. 159.

35. 9/24/160. See also Hankey to Liddell Hart 21 December 1949 and 21 January 1951; and R. T. Paget file 1948–51 *passim.* R. T. Paget *Manstein: His Campaigns and His Trial* (Collins 1951) Foreword by Lord Hankey—see especially Chapter 13. Lord Hankey *Politics, Trials and Errors* (Pen In Hand, Oxford: Regnery, Chicago 1950).

36. Paget to Liddell Hart 1 August 1952 and 13 January 1953. 9/24/161 Liddell Hart's correspondence with *Manchester Guardian* March 1951.

37. Liddell Hart to Paget 7 March and Paget to Liddell Hart 13 March 1951. See also Liddell Hart to Hankey 26 April 1950, 7 and 13 March 1951.

38. 11/1950/7 'Talk with General de Lattre—lunch at Claridges' 2 April.

39. 9/24/162 *passim.*

40. 11/1952/8 'Note on Talk with the German Chancellor' 9 June. Liddell Hart to Hankey 10 June 1952.

41. See the comment by Major Kenneth Macksey in 'Liddell Hart: the Captain Who Taught Generals' *The Listener* 28 December 1972, p. 895.

42. 11/1950/8 'Shinwell—1½ hours Talk' holograph note, 15 May. On 20 June 1951 Lord Pakenham introduced Liddell Hart to an up-and-coming Labour M.P. 'young James Callaghan' (see 11/1951/5).

43. For typical references to his influence on Guderian, Rommel and other German generals see 11/1951/19–22. His influence on German and Israeli military thought and practice is discussed in Chapters 8 and 9 of this book.

44. 11/1951/5 Diary notes 20 and 23 June.

45. 11/1951/14 'A Reflection' November. See also ibid. 11 and 15. 11/1954/2 'Memorandum for Dorman-Smith' 2 January.

46. 11/1952/2 'A Reflection on Losing Interest in a Subject' 14 February.

47. 11/1955/9 'A Reflection at Sixty' 31 October. Paget to Liddell Hart 16 December 1955.

7

Deterrent or Defence, 1950–60

In reviewing *Deterrent or Defence* (1960), the late Alastair Buchan alluded to the familiar pattern of personal development whereby the radical thinker or man of action, brilliantly iconoclastic in his youth, tends in later years to become increasingly conservative. Liddell Hart, in his view conformed to this pattern, 'but with a happy result':

> The great radical among the strategists of the inter-war years, the hammer of the complacent assumptions of the French and British general staffs . . . is now one of the most cogent critics of the strategic radicalism of today. The daring thinker of the thirties has become in the sixties the apostle of common sense, cautious and sceptical of the equally facile notions that deterrence is a substitute for defence, that nuclear weapons have made war impossible, or that they are a substitute for men at arms.[1]

The quality which, in Buchan's view, distinguished Liddell Hart's writing from that of the host of economists, journalists, politicians and historians now crowding into the field of strategic analysis was 'a profound understanding of the realities of war, and, above all, of tactics'.

Praise of the book and its author was shortly forthcoming from an even more prestigious quarter. Senator (soon to be President) John F. Kennedy agreed to write a lengthy review of *Deterrent or Defence* for the *Saturday Review*, and in effect used it as a text on which to expound his own ideas on the West's major defence problems.

> No expert on military affairs [wrote Senator Kennedy] has better earned the right to respectful attention than B. H. Liddell Hart. For two generations he has brought to the problems of war and peace a rare imaginative insight. His predictions and warnings have often proved correct.[2]

On the fundamental issue, that responsible leaders in the West 'will not and should not deal with limited aggression by unlimited weapons whose use could only be suicidal', Kennedy remarked that he shared 'Captain Hart's judgement', and he differed only in being more optimistic as to the possibility of agreement with the Russians on effective means of arms control.

Though, like *Defence of the West*, *Deterrent or Defence* was composed mainly of articles previously published elsewhere—some as long ago as 1952—it was better organized and less repetitive than the earlier book, thus providing a trenchant exposition of Liddell Hart's thinking about the major issues of defence in the nuclear age. It deserved the praise which the majority of reviewers gave it and constituted a fitting conclusion to Liddell Hart's studies of contemporary strategy.[3]

Only the landmarks of the preparations for the defence of Western Europe against the expansion of the Soviet Union need be mentioned by way of introduction. NATO was created in April 1949 in response to the Soviet blockade of West Berlin. Eight of its original twelve members were on the European mainland—France, Belgium, the Netherlands, Luxembourg, Italy, Portugal, Denmark and Norway—and the others were the United States, Britain, Iceland and Canada. Greece and Turkey joined the alliance in 1952 and Federal Germany in 1955. When the Korean war broke out in 1950 it was widely feared that this was a feint contrived by the Soviet Union to distract attention from Western Europe where the real offensive would be launched. In these circumstances the American commitment to come to the aid of her NATO allies in event of attack seemed inadequate: what was needed was a permanent American military commitment to an integrated force with common supply lines and an international headquarters. The latter, Supreme Headquarters for the Allied Powers in Europe (SHAPE), came into existence in April 1951, but the land forces have never been fully integrated. In the early 1950s it still made sense to envisage the defence of Western Europe almost completely in terms of the Second World War, because although the Soviet Union had exploded her first atomic bomb in 1949 it would be several years before she possessed a nuclear arsenal. It is moreover extremely doubtful whether the United States had any atomic bombs in the years immediately after the end of the Second World War and even by 1950 she probably had no more than twenty. At a NATO conference at Lisbon in February 1952 the enormous target was set of ninety-six divisions, but it soon became apparent that nothing like this figure would ever be achieved. In the first place the European members, then in the throes of economic reconstruction, were reluctant to raise the necessary manpower. Sir John Slessor, as Chief of Air Staff, was one of the first to advance the adoption of a policy of 'massive retaliation' in order to reduce defence expenditure. But, more important, as America—under similar financial pressures—became increasingly committed, in the mid-1950s, to a policy of 'massive retaliation' with nuclear weapons, it was hard to see the point of creating an effective 'shield' of ground forces. Indeed to do so would be to increase the chances of making Europe a nuclear battlefield. These fears were compounded by suggestions that the American contribution to the ground defences on the Central European front should take the form of a 'tripwire' whose function was not so much an actual defence of the area as a trigger to set off an American nuclear attack in the event of any Soviet encroachment on Western Europe.

By 1957 this 'tripwire' doctrine had been killed by the impressive growth of Russian nuclear capability. The Russians exploded their first hydrogen bomb device in August 1953, much sooner than expected, and exactly four years later successfully launched an inter-continental ballistic missile (ICBM). Western scepticism about this achievement was banished by the successful launching of the Russian moon rocket on 2 January 1959. Thus within a decade the 'unthinkable' had happened and Russia had actually taken the lead in a vital aspect of military science.

These developments underlined the inadequacy of NATO's forces and doctrine. Little progress had been made by the mid-1950s in building up conventional forces in the Central Area, while reliance on the threat of American nuclear retaliation against a Russian conventional attack now lost what little credibility it had had before the latter became a fully-fledged nuclear power. The Korean war should have ended the brief period of nuclear innocence because it revealed the psychological and moral obstacles to the use of nuclear weapons in a conventional war.

Consequently in 1954 NATO adopted a compromise solution: the target in divisions was reduced to thirty but they were to be put in a high state of readiness and all were eventually to be armed with tactical nuclear weapons. The 'tripwire' concept was dropped in favour of providing a limited range of alternatives between surrender and total war. The introduction of tactical nuclear weapons in turn raised a host of practical and theoretical problems which were to trouble NATO planners and critics throughout the 1960s.[4]

Such, in barest outline, was the background to the strategic problems treated by Liddell Hart in *Deterrent or Defence*. In the 1950s he continued his life-long habit of maintaining a vast correspondence as a means both of gathering information and also of refining his own ideas. In the period under review, his most important 'touchstones' were Air Marshal Sir John Slessor (Chief of the Air Staff 1950–52), Rear-Admiral Sir Anthony Buzzard (Director of Naval Intelligence 1951–54), Professor P. M. S. Blackett (pioneer of operational research and a member of Sir Henry Tizard's team during the Second World War), Alastair Buchan (first Director of the Institute for Strategic Studies from 1958) and Brigadier (later Field Marshal Sir) Michael Carver (Director of Plans at the War Office 1958–59). He was a founder-member of the prestigious Military Commentators Circle, set up in 1954, and was closely associated with the group (which included Buzzard, Blackett, Denis Healey and Richard Goold-Adams) who established the Institute for Strategic Studies four years later.

Thus in Britain alone Liddell Hart's preoccupations in the defence field were now widely shared by men of outstanding ability in the Services, journalism and—to a lesser extent as yet—the universities, so that it would be difficult to show that he was original on the broad issues. Nevertheless an important tribute by a leading American strategic analyst should be noted. In 1957 Bernard Brodie, then a senior member of the RAND corporation, wrote that he was glad to mention, in a book review, 'the fact that you had led all the rest of us in advocating the principle of limited war ... I became in effect a follower of yours early in 1952, when I learned ... that a thermonuclear weapon would be tested in the following autumn and would probably be successful'.[5]

One of Liddell Hart's basic assumptions, already emphasized in *Defence of the West*, was that Russia would only undertake a military offensive against Western Europe in exceptionally favourable circumstances. There was no need to underline Russia's terrible experience between 1941 and 1945, and her post-war policy, though suspicious and diplomatically aggressive, could reasonably be interpreted as stemming from an obsessive concern with security. Imagining himself in the shoes of the Chief of the Soviet General Staff in 1952, Liddell Hart counselled Stalin against risking a major war unless he could be certain of annihilating Britain with a knock-out blow and paralysing any counter-action by the United States for a lengthy period. It would be wiser to pursue a camouflaged warfare policy and strategy—'what our opponents call "the cold war". ... As a Soviet soldier', he concluded in this fictitious analysis, 'I am profoundly conscious of the fundamental truth of Lenin's maxim ... that "the soundest strategy in war is to postpone operations until the moral disintegration of the enemy renders the mortal blow both possible and easy"'.[6]

Later he was to suggest, and of course the truth is still uncertain, that in the immediate post-war years, when she possessed an overwhelming superiority in conventional forces, Russia had been deterred less by the threat of America's comparatively few atom bombs than the thought of her troops gaining first-hand knowledge of the material prosperity and liberty enjoyed by the citizens of Western Europe. 'They [the Russians] were frightened of their own people, and the evidence suggested that they were making a great effort to prevent them mixing and seeing the contrast between conditions in the East and the West.' Fuller shared Liddell Hart's belief that the 'cold war' was being deliberately employed to avoid 'hot war'. Russia's objective, he believed, was West Germany. 'Further, should it come to actual war, ... she will avoid risky experiments.'[7]

As early as April 1954 Liddell Hart penned a devastating critique of the American 'New Look' policy announced by Secretary of State John Foster Dulles at the beginning of that year. The basis of the policy was that 'local defence must be reinforced by the further deterrent of massive retaliatory power'; henceforth the West would depend primarily 'on a great capacity to retaliate by means and at places of our choosing'. Furthermore, Vice-President Nixon was reported as stating in March 1954: 'We have adopted a new principle. Rather than let the Communists nibble us to death all over the world in little wars, we will rely in future on massive mobile retaliatory powers.'

Liddell Hart was well aware that the new policy had been prompted by the desperate need to contain Communist expansion at greatly reduced cost, but in his view it attempted to reconcile three irreconcilable ideas:

> The concept of a quick knock-out punch by air delivery of atomic bombs in the heart of Russia.
>
> The concept of countering invasion by Russia's vastly larger land forces through the use, on the battle-front of the new tactical atomic bombs and shells.
>
> The concept of containment without conflict—by deploying sufficient strength on the ground to provide an effective deterrent to aggression.

For Liddell Hart, however, the overriding argument against the 'New Look' policy was that it rested on an illusion: would any responsible Government, when it came to the point, dare to *use* the H-bomb as an answer to local and limited aggression? It would be a lunatic action to take the lead in unleashing this menace which, he feared, could well result in 'general suicide and the end of civilisation'.[8]

He also believed that 'to the extent the H-bomb reduces the likelihood of full-scale war, it *increases* the possibilities of limited war pursued by widespread local aggression':

> The aggression might be at limited tempo—a gradual process of encroachment. It might be of limited depth but fast tempo—small bites quickly made, and as quickly followed by offers to negotiate. It might be of limited density—a multiple infiltration by particles so small that they formed an intangible vapour.

He accepted that since Russia possessed the H-bomb and the means of delivery the West could not risk unilateral nuclear disarmament. But he urged the case for a minimum-size deterrent—the vast Strategic Air Force of World War Two was now obsolete. 'Quality matters more than quantity and a relatively small number of super-performance aircraft would provide, inherently, a stronger guarantee of reaching the target—if that became necessary.' Once this viewpoint was accepted by governments, the way would be open to provide the conventional forces which really were needed and at an acceptable cost. These remarks may seem innocuous, unoriginal and even commonplace unless it is recalled that in the mid-1950s hair-raising notions were still prevalent in the Pentagon, at SHAPE and at Strategic Air Command (SAC) to the effect that if the bombers of the U.S. Strategic Air Command were unleashed, they could annul Russia's power of nuclear attack within a few days. Liddell Hart's conviction that there could be no 'winner' in a nuclear war, and that the only hope of preserving Europe lay in prevention, though far from novel, was still worth emphasizing.[9]

If the threat of massive retaliation was intended as a coherent strategic doctrine it was certainly vulnerable to Liddell Hart's criticisms. Michael Howard has suggested, however, that the doctrine may reasonably be seen 'as a political expedient—or even as a diplomatic communication, itself a manoeuvre in a politico-military strategy of deterrence. By these criteria the policy must be pronounced not ineffective.' Even so the weaknesses of the doctrine were obvious: American monopoly of nuclear weapons had not deterred the Korean conflict, and in 1954 superior nuclear power proved irrelevant to the outcome of the war in Indo-China. But what decisively undermined the doctrine's credibility at the end of the 1950s was America's new vulnerability to a pre-emptive or a retaliatory strike by Russian inter-continental bombers and missiles. With this alarming development, a new phrase entered the jargon of deterrence theory; namely the need for 'second-strike capability'. Henceforth the doctrine of deterrence would be critically dependent on technical calculations 'which stretched far beyond the orthodox boundaries of strategic thinking', and on which it was very difficult for independent analysts like Liddell Hart to comment with any authority.[10]

In trying to envisage what a European conflict between the nuclear powers would be like, Liddell Hart had changed his mind completely on one important aspect since the publication of *Defence of the West*, though this was legitimate in view of the transformation of the military balance as between Russia and America and the stock-piling of nuclear weapons by both sides. Whereas in the former book he had severely criticized the prevalent view that an atomic war was bound to be short, now in *Deterrent or Defence* he castigated Western strategists for basing their exercises and war games on the outmoded notion of a prolonged war. 'In all solemnity they try to work out the course of operations from H + 30 [days], H + 60, H + 90. It is astonishing evidence of the persisting influence of habit – but it makes no sense!' Field Marshal Montgomery was a notable offender in Liddell Hart's eyes for giving a lecture ominously entitled 'A Look Through a Window at World War III' in which he repeatedly used traditional terms such as 'win the battle' and 'win the war' and talked of 'bringing the war to a successful conclusion'. By contrast Liddell Hart cited with approval 'one of the most acute-minded air chiefs' (probably Slessor), who had shattered the complacency of a group of admirals and generals by remarking: 'It is no use trying to plan anything beyond the first six hours of another war.'[11] Unlike Herman Kahn, Liddell Hart was unwilling to 'think the unthinkable': for him any use of strategic nuclear weapons would signify the abandonment of reason.

Like other Western strategic commentators, by 1960 Liddell Hart was perplexed by the increasing dependence of the NATO forces on tactical nuclear weapons. In principle he conceded that there was much to be said for a policy of 'graduated deterrence'; that is, to reserve the use of H-bombs as a final resort while applying the minimum force necessary to repel any particular act of aggression, and to keep the conflict to the tactical level by striking against ground forces rather than cities. However he doubted whether escalation to all-out nuclear war could be avoided unless the use of tactical nuclear weapons could be confined to the immediate battlefield. In a chapter entitled 'Are Small Atomic Weapons the Answer?' he carefully weighed their advantages and disadvantages before concluding that the risks were greater than the benefits. He was not convinced that tactical atomic weapons favoured the defence, but a wider and more disturbing consideration was that control would be more and more decentralized until infantry battalions would have them. In theory, he concluded:

> These small-yield weapons offer a better chance of confining nuclear action to the battle-zone, and thus limiting its scale and scope of destructiveness—to the benefit of humanity and the preservation of civilisation. But once any kind of nuclear weapon is actually used, it could all too easily spread by rapid degrees, and lead to all-out nuclear war. The lessons of experience about the emotional impulses of men at war are much less comforting than the theory—the tactical theory which has led to the development of these weapons.[12]

Here Liddell Hart was in opposition to the conventional military wisdom of the later 1950s and early 1960s. Brigadier Michael Carver, for example, argued that tactical atomic weapons were an essential part of the deterrent and formed 'the real link between the shield and the deterrent'. Liddell Hart retorted that they were 'white elephants' since they could not be used. He was in distinguished company: Bernard Brodie had developed strong doubts about the wisdom of allowing NATO's forces to become dependent on tactical nuclear weapons, while Thomas Schelling suggested that the break between conventional and nuclear weapons was one of the rare 'natural' distinctions which made tacit bargaining possible in limiting war. The controversy over the role of tactical nuclear weapons was to continue throughout the 1960s, with the obvious risks balanced against the unpopularity and expense of large conventional forces. As Michael Howard wryly remarks: 'Only the sheer exhaustion of the participants keeps it from continuing still.'[13]

Behind the controversy over the introduction of tactical nuclear weapons lay an even larger issue: was a limited war with conventional weapons still possible in Europe in the nuclear age? Those who thought like Brigadier Carver put their trust wholly in deterrence and doubted the possibility of limited war, whereas Liddell Hart was convinced that it *could* occur if only NATO had the appropriate kind of conventional forces and tactical nuclear weapons remained subject to rigorous political control. Hence the aptness of his book's title *Deterrent* or *Defence*. In a chapter entitled 'Can NATO Protect us Today?' (and the situation has still not changed radically in the mid-1970s), he had no difficulty in showing the pathetic deficiencies and defects of NATO's ground forces:

> The existing divisions vary in size, mobility, equipment and

> weapon-power. The time required for mobilisation in the different national armies varies from three to forty-five days. The difference in types of weapon, types of vehicles, and supply systems are a great handicap to strategic flexibility—making it difficult to move divisions from one sector to another. It is hard to imagine such a mixed force maintaining a prolonged resistance, if it were also heavily outnumbered.

The basic truth, as he correctly perceived, was that ever since the late 1940s Free Europe's safety had depended on the deterrent effect of Strategic Air Command—its capacity to answer any Russian aggression on the ground by retaliating with nuclear bombs against the Russian homeland. The huge outlay on NATO's ground forces and tactical air force had made little or no contribution to the deterrent.[14] Yet by conceding so much he was, in effect, endorsing the doctrine of massive retaliation. Liddell Hart's major contribution to the problem in *Deterrent or Defence* was to describe in detail the organization, weapons and tactics which would render NATO forces capable of fighting a limited conventional war. It was an important part of his thesis that since the Napoleonic Wars ever fewer defenders were needed to hold a given length of front. Extrapolating from the German experience in the Second World War, when even with command of the air the Allies had needed a superiority of five to one to break their lines, he argued that the West could easily raise enough divisions to defend NATO's central front against the strongest forces that the Russians could deploy.

To deal with limited forms of aggression—frontier 'bites' and internal outbreaks fomented from outside—Liddell Hart proposed the creation of an extensive gendarmerie backed by mobile forces of high efficiency and in a state of constant readiness—like fire brigades. A relatively small professional Army would be much better suited than a short-service conscript Army to such a 'fire-fighting' role. It could be usefully supplemented by a 'superior militia type force, locally based'. In this chapter he *did* accept that tactics, movement, formations and organizations must be adaptable to the possibility that nuclear weapons might suddenly be used. He envisaged a 'new model' Army consisting of two types of active troops:

> The striking element would consist of a number of handy-sized

> armoured divisions, mounted entirely in cross-country vehicles that can move off the road. They would be trained to operate in controlled dispersion like a swarm of hornets, offering little target to a nuclear bomb or missile if such were used.
>
> The other type, for policing and for mobile defence, would be 'light infantry' divisions. They would also be completely capable of moving off the roads—but not through mechanisation. Their cross-country capacity would come from lightness of equipment.

More original was his idea of additional insurance to be provided by militia-type forces (a superior form of Home Guard), organized to fight in their own locality and to maintain themselves from local stores distributed in underground shelters. Such forces would be economical as regards transport; they would provide less of a target than regular formations; they would be less liable to interception; and would become effective with far shorter training.[15]

These proposals were remarkably consistent with Liddell Hart's earlier conceptions of a 'new model' Army (except that he was now more concerned with infantry than tanks), and with his general outlook on war. Nuclear weapons only served to reinforce the view he had maintained throughout the Second World War; namely that 'to aim at winning a war, to take victory as your object, is no more than a state of lunacy'. There was no sense in even planning for such a war: 'the destruction and chaos would be so great *within a few hours* that the war could not continue in any *organised* sense'.

In reviewing the book, Michael Howard noted that there was a gap in Liddell Hart's thinking at this point, for if a purely conventional defensive system *did* succeed in repelling a conventional attack and nuclear war was not unleashed, there was at least a logical possibility that a lengthy war of the Korean type might ensue.[16] Since NATO countries then and ever since have proved unwilling to raise the kinds of conventional forces and reserves that Liddell Hart proposed, we must be grateful that these ideas have not been put to the test.

Another aspect of defence on which Liddell Hart was remarkably consistent in his thinking was his advocacy of gas which, in *Deterrent or Defence*, he described as a more hopeful alternative to tactical atomic weapons as an effective but non-suicidal means of defence. He was surely correct in asserting that the failure to make much use of gas in warfare since 1918 was due to professional conservatism as well as to fear of retaliation. Since he had himself been badly gassed at the Somme in 1916 he had a particular right to his opinions on this highly emotional issue, and also constituted an example of his point that gas casualties had a far better chance of living an active life and reaching old age than those wounded by shellfire or, far worse, nuclear weapons. He called attention to the development of non-lethal gases which, by the end of World War II, could penetrate advancing tanks and knock out their crews in a matter of seconds. While mustard gas was not so decisive as an attack stopper, it possessed immense disabling and delaying power:

> It is particularly absurd to forego the defensive use of mustard gas, the most obstructive yet least lethal of weapons, while adopting the use of nuclear weapons—which are weapons of mass-slaughter, and violate the lawful code of warfare on more counts than such a weapon as mustard gas, which is relatively humane.

He also referred frequently, in print and in conversation, to a new nerve gas which in experiment had had the effect of making a cat terrified of a mouse.[17] Truly the perfect weapon for the humane philosopher of limited war!

In speculating why NATO's forces (or, for that matter, the Soviet forces also) have not adopted gas as a standard tactical weapon one must first concede that the experience of the kind of gases used in the First World War, particularly chlorine and phosgene, left a powerful legacy of revulsion and horror—feelings which were revived when the Italians employed gas in the Abyssinian war. In addition to the emotional barrier, however, two important arguments against gas may be discerned. The first relates to its practicality at the tactical level. There had been several unfortunate occasions in the First World War when gas either had been completely ineffective or had actually wafted back among the user's attacking troops. Professor Blackett was one of Liddell Hart's critics on this topic, suggesting that gas was too dependent on weather conditions and, moreover, since the attacker could choose the optimum conditions, it would actually handicap the defender—presumed to be the Soviet Union and the West respectively. Brigadier Carver also found Liddell Hart's arguments unconvincing:

historical experience suggested to him that antidotes would quickly be found and he did not believe that gas could be used to achieve surprise on a significant scale. On the other hand it should be mentioned that General Fuller completely endorsed Liddell Hart's thinking on this subject.[18] A greater obstacle, however, was simply the knowledge that gas (and chemical weapons generally) constituted a frightening Pandora's Box from which, once opened, agents truly horrific in their uncontrollable and long-lasting effects might emerge. Liddell Hart did not squarely face the problem that escalation was just as likely in gas and chemical as in nuclear warfare.

As Alastair Buchan pointed out in the review quoted at the beginning of this chapter, on the great strategic issues of the 1950s Liddell Hart could hardly be considered an *enfant terrible*, since his critical viewpoint was for the most part commonsensical and widely shared. His unconventionality is more evident in his attitudes to guerrilla warfare and passive resistance. Here his outlook is, at first glance, very surprising. It might have been expected that, as an admirer of T. E. Lawrence and the proponent of the strategy of indirect approach, he would have been sympathetic towards guerrilla operations, but such was not the case. Already, in *Defence of the West*, he had answered the question 'Were we wise to foster "Resistance Movements"?' (in the Second World War) with a definite negative. To question the wisdom of Churchill's policy of fomenting partisan activity in Nazi-occupied Europe was, he recalled, to appear lacking in resolution, and almost unpatriotic—yet he had done so. While he did not deny that the armed Resistance forces had imposed a considerable strain on the Germans, he doubted whether the effort had been worthwhile on a wider view. He believed that the effect of guerrilla operations was largely in proportion to the extent to which they were combined with the operations of a strong regular army that was engaging the enemy's front, and drawing off his reserves. 'They rarely became more than a nuisance unless they coincided with the fact, or imminent threat, of a powerful offensive that absorbed the enemy's main attention.' (T. E. Lawrence had of course exploited this advantage.) In viewing guerrilla activity first and foremost from the standpoint of the regular officer his attitude was remarkably similar to that of his *bête noire*, Clausewitz. As we have seen in other contexts, Liddell Hart's sympathies were with regular armies and with the honourable professional traditions which flourished in the Continental officers' corps. His sympathies clearly lay with the German occupation forces rather than the partisans in the Second World War. Partisan activity—and he did not distinguish between different types of irregular warfare—was often counter-productive in that it provoked reprisals much severer than the injury inflicted on the enemy: 'The material damage that the guerrillas produced directly, and indirectly in the course of reprisals, caused much suffering among their own people, and ultimately became a handicap on recovery after liberation.'

His most serious objection was, however, on moral grounds. Armed resistance attracted not only lovers of freedom but also many 'bad hats', who received licence to indulge their vices and work off grudges under the cloak of patriotism.

> Worse still was its wider amoral effect on the younger generation as a whole. It taught them to defy authority and break the rules of civic morality in the fight against the occupying forces. This left a disrespect for 'law and order' that inevitably continued after the invaders had gone.

By contrast Liddell Hart was impressed by the evidence emerging from countries such as Norway, Denmark and Holland after the Second World War, that non-violent resistance had been more effective, besides lacking long-term ill effects. The Nazis knew how to deal with violence, but were baffled by subtler forms of opposition.[19]

In *Deterrent or Defence* Liddell Hart gave sympathetic treatment to the truly radical ideas expressed in Sir Stephen King-Hall's book *Defence in the Nuclear Age*. The latter boldly followed opposition to nuclear weapons to its logical conclusion: not only should Britain unilaterally abandon 'the deterrent', but she should also cease to rely on American nuclear cover. To meet the—admittedly remote—possibility of Russian invasion, King-Hall proposed a defence policy of non-violent resistance. This was not to be passive resistance in the literal sense—as practised by Gandhi against the British—but was to embrace political warfare waged in an offensive spirit designed to drive home to the Communist peoples the advantages of the Western way of life.

Liddell Hart praised King-Hall's policy on moral grounds and did not entirely discount its practicality. He rightly pointed out, however, that non-violent resistance had only tended to work against opponents whose code of morality was fundamentally similar (and, he might have added, in the case of the British in India some sympathy with the resisters). Stalin was hardly vulnerable to such tactics, while they had only excited Hitler's impulse to trample on contemptible weakness. Nevertheless Liddell Hart retained two serious doubts about non-violent resistance as a national policy, as distinct from that of a cohesive religious or political movement. The first was whether the nation as a whole, or any likely Government, could be persuaded to embark on such a revolutionary experiment. The second was whether such a policy, if adopted, could possibly succeed—since the effectiveness of non-violent resistance is undermined if even a small proportion of the community play into the opponent's hands—through weakness, self-interest or pugnacity.

> Such instincts tend to be much more prevalent in a nation than in a sectional, and spiritual, movement. Comparatively, an army is more dependent on its strongest elements, while an unarmed force is more dependent on its weakest elements. To make non-violent resistance a national affair is an extremely difficult task: probably the most important thing to do is to educate people and convince them that it is a workable policy.[20]

In an expanded version of the essay on guerrilla warfare and non-violent resistance published in 1967, and quoted above, Liddell Hart provided a good account of the development of modern guerrilla warfare from T. E. Lawrence and Mao to the contemporary campaigns in South-East Asia, Algeria, Cyprus and Cuba. He argued that a key factor was the ratio of space to forces. In the past, he suggested, guerrilla warfare was a weapon of the weaker side and thus primarily defensive, but in the atomic age it might be increasingly developed as a form of aggression suited to exploit a situation of nuclear stalemate. Curiously, however, he attributed the vast extension of subversive warfare in the last twenty years largely to the Churchillian policy of instigating and fomenting popular revolt in Nazi- and, later, Japanese-occupied countries. This was to ignore the long history of guerrilla operations against imperial rule in the nineteenth century which had gathered impetus in many countries after the First World War. Liddell Hart was well aware that Mao Tse-tung's ultimately successful guerrilla operations had begun in the late 1920s. The Japanese conquest of a large part of the British, French and Dutch Far Eastern empires and their brutal treatment of the inhabitants did far more to foment revolutionary warfare than the defeated imperialist regimes, who became the targets after 1945. One can only attribute this distortion of history to his extreme bitterness towards Churchill.

Liddell Hart's discussion of guerrilla warfare was altogether too narrow and idiosyncratic in that he appeared to be thinking mainly of the established nation states overrun by the Nazis in the Second World War. A country such as Denmark could hope for eventual liberation by the regular armies of the Allies with a minimum of self-help in the form of guerrilla activities, but this option was not available to subject peoples seeking national autonomy not only from the Japanese invader but also from their British, French or Dutch imperial Masters. Liddell Hart's analysis was also idiosyncratic in that he largely ignored the fierce nationalism which provided guerrilla warfare with its revolutionary dynamism. Leaders such as Ho Chi Minh, Giap, Grivas and Castro were probably not much concerned with the possible long-term ill effects of their campaigns compared with their determination to expel foreign occupation forces and achieve national independence.[21]

In his essay published in 1967, in a volume entitled *The Strategy of Civilian Defence*, Liddell Hart added an interesting section on the 'Strategy of Resistance' in which he characteristically advocated indirect methods rather than direct strike action and blunt refusal to carry out the orders of the occupying power. He made no claim of originality for the idea that non-co-operation should be concealed behind apparent acquiescence. Such a strategy would become even more baffling if practised with a cheerful smile suggesting that mistakes were due to incomprehension or clumsiness; in other words the anti-heroic methods of the legendary Good Soldier Schweik. In Liddell Hart's view this subtle kind of resistance 'cannot really be dealt with in terms of force; indeed nothing can deal with it. There is really no answer to such go-slow tactics'.

He summed up the essential aims of the strategy of civilian resistance as follows:

> The more the occupying forces can be made to spread, the more complex their problems become. That, I would say, should be the guiding principle in planning civilian defence, and an attempt should be made to adopt strategies and methods which cause maximum strain and therefore overstretch. As in guerrilla war, so in civilian defence, the principle applies that one should aim for a multiplicity of offensive actions—offensive in the psychological sense, and coupled with multiplicity of human contact with the occupying forces.

By this last phrase he had in mind a precept already successfully practised by Mao Tse-tung with his idea that prisoners should be converted to Communism by considerate treatment. Whereas violent forms of resistance tend to assume that all members of the occupying forces are enemies, civilian defence should regard them as fellow human beings and get this human feeling spreading, thus weakening the opposition's sense of solidarity and authority. He admitted that there was a long way to go before any government would see the relevance of the idea of civilian defence, though he had found some awareness of its possibilities in the Scandinavian countries. 'Probably the more that governments come to realize their incapacity for effective military defence, the more they will begin to take non-violent civilian defence seriously.'[22]

In asking some questions about the essays collectively published in *The Strategy of Civilian Defence*, Professor Thomas Schelling made two penetrating comments which were pertinent to Liddell Hart's contribution. First, he pointed out that it was comparatively easy to make the case for the success of non-violent civilian resistance as a form of protest or as a bargaining technique that leads to amelioration, compromise and accommodation. But the book was not about civilian resistance as a method of protest, as a method of alleviating conditions, as a method of denying a conqueror *some* of what he wants and of coming to terms with him. Rather the book sought to establish the much harder case that this strategy could actually *win* in the sense that it could make a conqueror withdraw or dissuade him from conquest in advance. Schelling doubted, and Liddell Hart probably would have conceded the point, whether non-violent defence by itself could actually make a tyrant retreat or withdraw 'at least within the time span that the word "defence" suggests'.

Schelling's second and more serious question was: 'Is there any reason to suppose that the techniques of non-violent resistance are more available to good people than to bad, to right causes than to wrong, to democrats than to demagogues, or to defence rather than offence?' Liddell Hart, for example, envisaged 'the enemy' as an alien occupying power, but could not the same techniques be used against a regime desperately trying to become democratic in a country with a military or authoritarian tradition? Which target was more vulnerable: the Nazi regime or the Weimar Republic? The segregated South or the Supreme Court? The University of Peking, or the University of California? 'It would be nice,' wrote Schelling, 'if all good things clustered together, and non-violence (which sounds so much better than violence) had a necessary affinity for the good and the right and the pure—for the meek and all those who should inherit the earth. Maybe it does, but we do not know it yet.'[23] A great deal of experience of popular and more-or-less non-violent protest movements since 1967 suggests that Schelling's scepticism was fully justified.

Deterrent or Defence concluded with a reflective epilogue in which Liddell Hart admitted that, in contrast to the 1930s, he now felt very dubious about attempting any predictions concerning the nature of a third World War, beyond the basic certainty that it would immediately produce chaos. While he welcomed the possibility of an agreement to limit the nuclear arms race as a step towards relieving tension, he felt that any such agreement would be inherently unstable because of the rapid rate of scientific development. There was a widespread feeling in the West that no co-existence or compromise was possible with the Communist regimes of Russia and China, but nothing could be more fatal than the feeling 'it's bound to come —let's get it over'. It would be wiser for everyone to become adjusted to a continuation of difficulties, and tense manoeuvring, than to look for any definite settlement. For him, history offered comforting evidence that 'no antagonistic line-up is fixed and final'. Tension as intense as during the last decade was almost bound to relax eventually if war was postponed long enough. Russia did not want war and the West could reassure her by developing NATO's conventional defensive shield rather than its apparently offensive nuclear capability.

Although some reviewers considered Liddell Hart's moralizing epilogue platitudinous, his practical advice was eminently sensible. Whether or not President Kennedy recalled this particular passage in the book, he certainly acted according to its spirit when confronted with the Cuba missile crisis in 1962:

> Study war and learn from its history. Keep strong, if possible. In any case, keep cool. Have unlimited patience. Never corner an opponent, and always assist him to save his face. Put yourself in his shoes—so as to see things through his eyes. Avoid self-righteousness like the devil—nothing is so self-blinding. Cure yourself of two commonly fatal delusions—the idea of victory and the idea that war cannot be limited.[24]

This and the previous chapter have attempted to show that Liddell Hart deserves a high ranking among the early strategists of the nuclear age such as Slessor, Brodie and Blackett. He had been among the first to appreciate the limitations of atomic deterrence in the 1940s, and in the following decade he was one of the severest critics of the policy of massive retaliation. Throughout the period he consistently stressed the need for the West to build up mobile conventional forces both to supplement the nuclear deterrent and to fight limited wars. Perhaps most important of all, in the sphere of grand strategy he urged the West's leaders to be patient, cautious and conciliatory. Such ideas were not so obvious and commonplace in the era of the Korean war and Senator McCarthy's witch hunt, but by the mid-1970s East–West relations (at least in external relations between the super-powers) seem to be going the way he predicted.

By the later 1950s as a freelance commentator Liddell Hart was competing at a grave disadvantage with a host of strategic analysts drawn from such varying disciplines as physics, engineering, mathematics and economics, many of them working in teams in defence research institutes such as the RAND corporation where they enjoyed access to classified information. Consequently, as Michael Howard rightly remarks, 'They analysed the technical problems of deterrence with an expertise which earlier works had naturally lacked.' When Robert McNamara became Secretary of Defense early in 1961 many of these 'defence intellectuals' were drawn into government and, according to Raymond Aron, 'lived their finest hour' before becoming somewhat discredited by the Vietnam war.[25]

Even had Liddell Hart been younger and with an assured private income, it is doubtful if he would have been inclined to adapt his mode of thinking to the new style of strategic analysis in which technical expertise, games theory, conflict analysis and theoretical models featured prominently. Despite his undeniable interest in operational research and 'scientific analysis', Liddell Hart was essentially a pragmatic thinker who drew his 'lessons' from historical experience (supplemented by a wide range of personal contacts), and who excelled at communicating complex military issues to a wide readership. The validity of the historical approach to contemporary defence problems was severely downgraded during the later 1950s and 1960s when theoreticians such as Herman Kahn enjoyed greater prestige—and influence—particularly in the United States. Whether justly or not, the failure of theorists to predict or prevent the débâcle in Vietnam has given the more traditional disciplines a new lease of life. As Raymond Aron asked rhetorically, in summing up his view of the development of modern strategic thought in 1969:

> Have the innovators got anything better to offer us to assure peace, to win the cold war or bring it to an end, to take just decisions or to change the nature of relations between states than prudence nurtured by history at least as much as by laboratory experiments on the psychological effects of confidence? Does make-believe teach us more about the reality of a situation or the development of a crisis than historical and sociological analysis?[26]

Both Aron's approach and his conclusion ('to guarantee that wars remain limited, is it not to accept permanent war?') would have met with Liddell Hart's endorsement.

Throughout the 1960s Liddell Hart continued to publish articles on most of the topics covered in *Deterrent or Defence* such as the problems of NATO and the ratio of force to space. He also commented on the great military events of the decade such as the Cuba missile crisis, the Suez operation and the Six Day

Arab–Israeli War. But *Deterrent or Defence* was really his swan song as an expert analyst of contemporary affairs, as he made clear in a letter to Alastair Buchan tendering his resignation from the Council of the Institute for Strategic Studies:

> For a long time I have found it difficult to devote time to history while striving to keep abreast of current affairs, and have come to realise that it would be better to concentrate my remaining energy on one or the other. In making a choice, it is apparent that history offers more interest and scope, while the possibility of effectively influencing defence policy here is slight, and the demand for my views and advice on current problems comes mainly from other countries—which is no adequate incentive to continue the effort. So it is not difficult to make the choice.[27]

Notes

1. Alastair Buchan in the *Observer* 3 July 1960. Surprisingly, Liddell Hart accepted Buchan's description and wondered if he might be becoming *too* mellow with advancing years, in which case a possible title for his memoirs might be 'The Taming of the Tiger': Liddell Hart to Buchan 4 July 1960.
2. *Saturday Review* 3 September 1960.
3. Although he continued to publish articles on contemporary affairs, *Deterrent or Defence* was his last attempt to formulate a general statement of his strategic ideas in a book.
4. A. Buchan 'Between Sword and Shield' in *Encounter* July 1959. R. Aron 'The Evolution of Modern Strategic Thought' in Institute for Strategic Studies *Adelphi Paper* no. 54 February 1969 pp. 2–3. *Deterrent or Defence* pp. 39–41, 133.
5. B. Brodie to Liddell Hart 26 April 1957; cp. Michael Howard 'The Classical Strategists' *Adelphi Paper* no. 54 pp. 20–21.
6. *Defence of the West* pp. 148–150. *Deterrent or Defence* pp. 5, 16.
7. Liddell Hart 'Lessons from Resistance Movements—Guerrilla and Non-Violent' in Adam Roberts (ed.) *The Strategy of Civilian Defence* (Faber 1967: *Civilian Resistance* ... Stackpole 1968) p. 208. Fuller to Liddell Hart 28 December 1952. By 'risky experiments' Fuller probably meant large-scale airborne operations which Liddell Hart had favoured in his guise of Russian Chief of Staff.
8. *Deterrent or Defence* pp. 21–23. See also 11/1955/5 'Reflection on the Current Defence Problem of the Western Powers' August.
9. *Deterrent or Defence* pp. 23–25. See also 11/1957/14 'Strategy in the Melting Pot' 16 January.
10. M. Howard 'The Classical Strategists' op. cit. pp. 22–24.

11. *Deterrent or Defence* pp. 55–57. See also P. M. S. Blackett *Atomic Weapons and East–West Relations* (C.U.P. 1956) pp. 16–17.

12. *Deterrent or Defence* pp. 58–61, 74–81.

13. Michael Carver to Liddell Hart 10 October and Liddell Hart to M. Carver 13 October 1958. M. Howard 'The Classical Strategists' p. 25.

14. M. Carver to Liddell Hart 21 October 1958. *Deterrent or Defence* pp. 134–136.

15. Ibid. pp. 62–67, 94–96.

16. M. Howard in *Survival* September–October 1960.

17. *Deterrent or Defence* pp. 62, 82–88.

18. P. M. S. Blackett to Liddell Hart 28 September and Liddell Hart to Blackett 1 October 1959. Personal conversation with General Fuller *c.* 1962.

19. *Defence of the West* pp. 53–57.

20. *Deterrent or Defence* pp. 218–221. Liddell Hart in *The Strategy of Civilian Defence* op. cit. p. 206.

21. Liddell Hart ibid. pp. 196–205.

22. Ibid. pp. 206–209.

23. Ibid. pp. 302–308.

24. *Deterrent or Defence* pp. 247–257.

25. M. Howard in *Adelphi Paper* no. 54 op. cit., pp. 20, 26; R. Aron ibid. p. 5.

26. Ibid. p. 17.

27. Liddell Hart to A. Buchan 15 August 1960.

8

Liddell Hart and the German Generals

Although it is often remarked that 'the pen is mightier than the sword', in practice it is difficult to demonstrate the former's precise effect on military events, and even harder to prove the influence of a particular 'pen' (or writer) on the 'sword' (or commander). Clausewitz offered one explanation in his wise remarks on the role of theory:

> It should educate the mind of the future leader in War, or rather guide him in his self-instruction, but not accompany him to the field of battle; just as a sensible tutor forms and enlightens the opening mind of a youth without, therefore, keeping him in leading strings all through his life.[1]

In other words, generals seldom fight a battle with a particular textbook in mind, and since they bear the chief responsibility for success or failure it is asking a lot to expect them to attribute their victory, if they secure it, to an intellectual influence. Moreover most intelligent commanders will have read widely in their formative years and with the best will in the world may find it impossible in retrospect to single out *one* author who decisively influenced their outlook.

Nevertheless after 1945 several outstanding German generals *were* prepared to acknowledge Liddell Hart as their mentor. The purpose of this chapter is to present the evidence for these assertions of influence and to describe the context in which they were made. Since this study relies heavily on Liddell Hart's archives, supplemented by the post-war memoirs and campaign histories of the German generals, it makes no claim to be exhaustive or definitive. There is clearly scope for specialized research in the German military journals and training manuals of the 1930s to determine, for example, the influence of other British theorists such as Generals Fuller and Martel who are sometimes mentioned without elaboration in the memoirs.

The first point to establish is that there is no doubt whatever concerning the general popularity of Liddell Hart's publications in the German Army in the 1920s and 1930s. Several of his books were translated by senior officers in influential positions; his articles were circulated in large numbers and—in the mid-1930s—Liddell Hart was pressed to advise the Reichswehr on mechanization and, failing that, to attend manoeuvres. He wisely declined these invitations.[2]

A sharp distinction needs to be drawn between Liddell Hart's general influence and his specific influence on particular aspects of the development of armoured forces. The former was due to the sheer volume of his writings; to the fact that he was reporting on exciting new developments in Britain in the later 1920s and early 1930s; and to his propounding of novel ideas. Dr Robert O'Neill in preparing his book *The German Army and the Nazi Party* encountered numerous German officers of the inter-war period who mentioned Liddell Hart as a general source of intellectual stimulus and probing comment. The very fact that the Liddell Hart phenomenon existed was in itself a major encouragement to the post-1919 disciples of Moltke and Schlieffen. The fact that he was making a major contribution to their training courses, exercises and mess discussions during a period of clandestine rearmament 'gave the whole intellectual development of a generation of German officers a lift forward—no doubt small but definitely perceptible'.[3]

Before providing a sample of the evidence for this general pervasive influence it is important to stress the continuity of the tradition of mobile operations, especially on the open plains of Eastern Europe. General Hans von Seeckt, Chief of the Army Command from 1920 to 1926, had after all been Chief of Staff to Mackensen during the preparation for the great break through at Gorlice-Tarnow in 1915. During his post-war period in command of the Army, Seeckt planned lightning strikes against the greatly superior Polish forces in event of war. Thus, though Seeckt was not a proponent of *blitzkrieg* in the technical sense, he had a shrewd idea of what he was looking for in terms of strategic mobility.[4]

Another aspect of this continuing tradition of mobility is a very pragmatic exploitation of whatever means happen to be at hand, and in this case of mechanized transport, to which the wide open spaces of Eastern Europe lent themselves. Thus, long before the term *blitzkrieg* was coined, and even before the first impressive use of tanks at Cambrai in November 1917, a German infantry force mounted in trucks executed a lightning strike as part of von Falkenhayn's offensive against Rumania in 1916. In 1920 the Polish forces made even more daring use of armoured cars to spearhead strategic raids by motorized forces. The first raid was in April 1920 in the Ukraine and the second in September against the railway junction of Kowel, well in the rear of the Soviet troops retreating from Warsaw. Both operations involved the employment of two truck-borne infantry battalions with some truck-drawn field guns; the latter witnessed an encircling sweep of about one hundred miles. These operations and post-war plans directed against Poland provided an important pre-conditioning to the later ideas of Guderian and others on mechanized operations. The significance of theorizing has perhaps been exaggerated, in comparison with pragmatic adaptation and practical experimentation, in tracing the development of German armoured forces.[5]

Ironically, by limiting the Reichswehr to 100,000 long-service soldiers and depriving it of tanks and heavy weapons, the Allies provided incentives for study, improvisation and secret experimentation. Nor should it be overlooked that it was Ludendorff who had first restored mobile warfare to the Western Front by employing revolutionary infiltration tactics to break through the Allies' defences in March 1918. This breakthrough was all the more remarkable in that it was achieved mainly by infantry and artillery, with only a few tanks playing a minor role. Germany had lagged behind in the development of armoured forces: the first tanks were not commissioned until 1917 and the first detachment of home-produced A7V tanks was ready in January 1918. The A7V was a 30-ton turretless vehicle with a 57mm gun and six machine guns, manned by a crew of eighteen. Only twenty were completed by the end of the war. A further eight units had been added by the Armistice, though six were equipped with captured British Mark IV tanks. German tanks took part in about a dozen actions, including the first tank-versus-tank combat at Villers-Bretonneux on 24 April 1918. Thirteen German tanks were employed as a spearhead and whenever they appeared the British line was broken.[6]

Thus the Armistice and the restrictions imposed by the Treaty of Versailles constituted only a temporary check on an

Army determined to profit from its mistakes by developing the tactics and instruments of mobility already in evidence in the latter part of the war. Finally, to these short-term incentives must be added the intellectual tradition of the now-forbidden General Staff. Schlieffen's plan may have failed in 1914 but the spirit of his doctrine survived. As Field Marshal von Manstein recalled after the Second World War:

> Schlieffen's theories were much studied in the German Army between the wars and had great influence on the strategical *and* tactical thinking. The idea to outflank and enclose the enemy governed the Germany strategy and tactics.[7]

Among the enthusiastic readers and distributors of Liddell Hart's publications were two of the leading pro-Nazi German generals, Blomberg and Reichenau. Liddell Hart met the former, who was shortly to become Hitler's Minister of War, at the Geneva Disarmament Conference in 1932. Blomberg had recently read Liddell Hart's biography of Marshal Foch and had been particularly impressed by its emphasis on the power of the defensive. Blomberg remarked that he was correcting the orthodox obsession with the offensive in his training—he was then commanding the most important district *Wehrkreis 1* in East Prussia. He confessed that this preoccupation with the offensive was perhaps a natural consequence of Germany's military situation but was also a deduction from her war experience.[8]

Later in 1932 Colonel (later Field Marshal) Reichenau, who was then Chief of Staff to the 1st Division, and in 1939 commanded the Army which broke through to Warsaw, informed Liddell Hart in a letter that he was translating *Foch* but was having difficulty in finding a German publisher. His chief interest in the book was its 'immanent value': 'Here I found a new and unusually high standard of judgement, not following obsolete theories, but setting new rules—if there are any at all'. Both officers were also responsible for translating *The British Way in Warfare* in 1932 and for circulating Liddell Hart's ideas on mechanization throughout the Reichswehr.[9]

By the early 1930s Liddell Hart was already aware from personal contacts that some of his writings were being avidly studied in Germany. In an autobiographical note, filed under the year 1931 but evidently written a few years later, he noted:

> A study of future warfare and scheme for gradual mechanisation, which I produced in 1921, was translated by the German Ministry of War, and circulated throughout the Reichswehr, and by other armies subsequently.[10]

Fuller reported to Liddell Hart in February 1935 that on a recent visit to Berlin he had seen several copies of the latter's *Lawrence* (in German) on the bookstalls 'and the Second in Command of the S.A. told me that he had ordered 5,000 copies of [Liddell Hart's book] *The Future of Infantry* in translation'.[11] Two years earlier, Sir Maurice Hankey, Secretary of the Committee of Imperial Defence, informed Liddell Hart:

> I heard from a high military authority in Berlin the other day that yours and Fuller's writings are widely read and eagerly awaited throughout the German army.

Hankey's informant was probably the British Military Attaché Colonel (later General Sir Andrew) Thorne, who subsequently wrote to confirm the point:

> I was (as you may remember) Military Attaché in Berlin from April 1932 to May 1935, and during that time there I could not fail to be impressed by the extent to which both Liddell Hart's and 'Boney' Fuller's books were being studied by officers of all ranks and arms in the German Army. I knew both Blomberg (Minister for War) and Reichenau (Chief of the Defence Staff) very well, and they were both engaged in translating books by these two authors for use for non-English speaking German officers. They frequently sought my help to elucidate some point which was not quite clear to them.
>
> Liddell Hart's themes and arguments had as stimulating an effect on the German military mentality as it had on ours, and I am not sure they even didn't get more value out of it than we did, as they were on the whole much better grounded in principles than we were.[12]

As a final example of his European popularity, Liddell Hart wrote to General Thorne in 1942 that he had recently been consulted by a refugee Czech officer, Colonel Miksche, who had bluntly told him that if he had merely been the best-known

British military writer it would have meant nothing to him, 'but for years past he had constantly seen my writings discussed in the German and other Central European military reviews, and for any military writer to gain a "Central European reputation" *was* a real guarantee of professional knowledge'.[13]

These quotations demonstrate that Liddell Hart's publications, including non-specialist works such as *Foch* and *Lawrence*, were translated and widely read in Germany. But it should not be overlooked that Fuller's name was equally well known to German readers and probably even more often mentioned in professional military journals such as *Wehrwissenschaftliche Rundschau* as an expert on mechanized warfare. Fuller, however, lacked some of Liddell Hart's skill with the pen; he did not have the platform of the press to publicize his views; and he did not make a systematic effort after the Second World War to collect evidence of his pre-war influence.[14]

It was, however, less on this general evidence that Liddell Hart founded his claim to have influenced the German theory and practice of *blitzkrieg* than on the specific acknowledgements of one man, General Heinz Guderian, who was credited by a fellow-general, who did not share his belief in Nazism, for '60% of what the German Panzer-waffe has become'. Another wrote: 'I would like to say once more how highly I value him [Guderian] as the creator and *master* teacher of our Armoured Forces, laying particular stress on the word "master".'[15] Guderian has provided a reasonably detailed account of the origin and growth of Germany's armoured forces before 1939 in his war memoirs *Panzer Leader*, and the same ground has been traversed in concise but scholarly style by Robert O'Neill with particular reference to Liddell Hart's influence.[16] There is no need therefore to repeat all the details here, merely to summarize Guderian's career and highlight the points at which Liddell Hart's writings were valuable to him.

Captain Guderian, a signals specialist with no previous experience of mechanical vehicles, first became interested in problems of mobility in 1922 when appointed a staff officer to the Inspector of Transport and employed in the Motorized Transport Department. His initial task was to devise methods of defence using motor transport to frustrate an invader's attempts at deep penetration, and he quickly realized that armour was essential for protection. A colleague, Lieutenant Volckheim, stimulated his interest in the use of armour in the Great War, and as early as 1923 or 1924 he came across the ideas of Fuller, Liddell Hart and Martel in the special periodical review *Wehrgedanken des Auslands*. Their writings implanted in his mind the crucial ideas that armour could be employed in the attack and that it must be organized as a self-contained spearhead rather than as a light reconnaissance force or an infantry support weapon. He later acknowledged the importance of this concept—several years before he even saw a tank—in the following terms:

> It was principally the books and articles of the Englishmen, Fuller, Liddell Hart and Martel, that excited my interest and gave me food for thought. These far-sighted soldiers were even then trying to make of the tank something more than just an infantry support weapon. They envisaged it in relationship to the growing motorisation of our age, and thus they became the pioneers of a new type of warfare on the largest scale.
>
> I learned from them the concentration of armour, as employed in the battle of Cambrai. Further, it was Liddell Hart who emphasised the use of armoured forces for long-range strokes, operations against the opposing army's communications, and also proposed a type of armoured division combining panzer and panzer-infantry units. Deeply impressed by these ideas I tried to develop them in a sense practicable for our own army. So I owe many suggestions of our further development to Captain Liddell Hart.[17]

By the mid-1920s Guderian was developing his ideas on mechanization in articles and lectures, but at first he encountered considerable scepticism and opposition. When, for example, he suggested that the primitive armoured troop-carriers then used for supply might be transformed into combat units, his inspector replied bluntly: 'To hell with combat! They're supposed to carry flour!' Opinion in the High Command became more favourable to the idea of tanks as a result of the first post-war manoeuvres in 1925 and 1927, the latter witnessing the first employment of dummy tanks on a large scale; and Seeckt gave cautious encouragement to the idea that tanks might acquire a distinctive role. Guderian was forced to clarify his thinking on tanks in 1928 when appointed to lecture to a knowledgeable audience of officers in the Motor Transport Section at Stettin; and in 1929 he saw his first 'live' tanks on a four-week visit to Sweden. By this time he had built up his own translation service with the specific purpose of reading and distributing new English books, articles and the early rudimentary training manuals. The latter provided the basic textbooks for German mechanized training during the next decade. At this stage Guderian was only one of several officers specializing in the development of tank tactics; unlike many of his colleagues, he did not have the opportunity to visit the secret Russian training centre at Kazan where officers were trained in the handling of tanks, early German models were tested, and comparisons made with foreign models.

In 1929 Guderian reached the definite conclusion that tanks working on their own or in conjunction with infantry could never achieve decisive results. Moreover, for tanks to make their full impact he realized that their support weapons would have to be brought up to their standard of speed and cross-country performance. Many senior officers remained sceptical: the Inspector of Transport Troops, General Otto von Stülpnagel, for example, forbade the theoretical employment of tanks in units larger than the regiment and dismissed the idea of Panzer divisions as a Utopian dream. It was precisely at this time, as a result of his study of Sherman, that Liddell Hart was formulating his concept of deep penetration carried out by armoured formations. Tanks were now harnessed to Liddell Hart's earlier concept of the 'expanding torrent' to achieve the essential condition of maintenance of momentum. As Dr O'Neill put it:

> For strategic penetration of hundreds of miles a concentration of a few divisions would be useless except against the weakest of foes. Whole armies of tanks were required.[18]

He goes on to suggest that in 1931–32 Guderian was in the right stage of development to absorb these ideas and to start turning them to practical effect. The latter realized that the only solution lay in the creation of Panzer divisions in which the spearhead of tanks would ideally be closely supported by equally mobile infantry, engineer, artillery and signals units. It seems likely that Guderian had read Liddell Hart's article 'The

Development of a "New Model" Army', which was published in the *Army Quarterly* in October 1924. Its most radical suggestion was that the new model division would be composed of three composite brigades of tanks, infantry in armoured carriers and self-propelled artillery. There would be 300 tanks in a division, including a battalion of medium tanks designed for the pursuit. Liddell Hart envisaged future land forces as composed principally of tanks and aircraft supported by a small force of siege artillery to smash defended strong points and of mechanically borne infantry to serve as 'land marines'—acting offensively and defensively. It was this latter emphasis on the need for close co-operation between tanks and infantry that most clearly distinguished Liddell Hart's concept from Fuller's 'all-tank' views which were followed by the Royal Tank Corps.[19]

While it seems reasonable to claim that Guderian was influenced by Liddell Hart in the 1920s on the vital point of the need for close co-operation between tanks and infantry, it would be straining the evidence too far to suggest that the first three Panzer divisions created in 1935 closely resembled Liddell Hart's 'New Model' division of a decade earlier. The resemblance was closest in the tank component. Indeed, as Richard Ogorkiewicz has pointed out, the earliest Panzer divisions did not differ much from some other contemporary developments such as the French Division Légère Mécanique (D.L.M.) and the British experimental Tank Brigade of 1934, each being based on a tank brigade backed by an infantry brigade. The British Tank Brigade actually provided a model for the first Panzer brigade. The latter consisted of two tank regiments each consisting of two tank battalions of Panzer I light tanks. This gave a nominal total of 561 tanks for the brigade—'a very powerful tank component, sufficient to satisfy the most ardent tank enthusiasts'.

It was in the supporting units and particularly the infantry component that the Panzer divisions differed from Liddell Hart's conception. He envisaged armoured units as consisting essentially of tanks supported by infantry in the form of 'land' or 'tank marines', whereas the Panzer divisions, in principle at any rate, possessed *as large a unit of infantrymen as of tanks*. In other words, whereas Liddell Hart pictured a small number of infantrymen being attached to a *tank* force, the Germans joined together large units of infantry, artillery and tanks to make a new type of formation which was not followed elsewhere. Thus the Germans managed to avoid both the pitfalls of the 'all-tank' theory of tanks working independently which prevailed in Britain, and the doctrine of tying the tanks to the support of infantry which became the practice in France. Richard Ogorkiewicz describes the salient characteristics of the original Panzer Divisions as follows:

> As well as having the tank brigade each panzer division also had from the very beginning a motorised infantry brigade whose role was to support and complement the tanks. The infantry, or rifle, brigade consisted of one 2-battalion truck-borne regiment and of one motor-cycle battalion. Each division also had a motorised artillery regiment with twenty-four 105mm. howitzers, an anti-tank battalion of towed 37mm. guns and an engineer company—quickly expanded into a battalion. There was also a reconnaissance battalion of armoured cars and motor-cyclists, as well as signal and divisional service units. The panzer division was thus a self-contained combined arms team in which tanks were backed by arms brought up, as far as possible, to the tanks' standards of mobility.[20]

Before the advent of Hitler and his rapid conversion to the notion of *blitzkrieg*, Guderian and his fellow tank enthusiasts probably encountered as much obstruction, half-heartedness and outright hostility as the British armoured school. In addition to the traditional 'horse and foot' supporters who disliked tanks *per se*, there was the more insidious opposition of those who wished to seduce tanks into support roles for the older arms. The cavalry wanted light mechanized forces to enhance its performance in reconnaissance and pursuit, while the infantry wanted tanks to pave the way for riflemen as they had done for the Allies in the First World War. Unlike his British equivalents, however, Guderian was fortunate in finding keen supporters in the highest positions in the Reichswehr; Blomberg, Reichenau and Fritsch in particular backed the development of armoured forces against the scepticism of more conservative officers such as the Chief of the General Staff, Beck. The latter is reported to have become so exasperated by the frequent references of Guderian and other tank enthusiasts to Liddell Hart that he wished for a change that he could have six months without hearing Liddell Hart's name.[21]

In October 1931 Guderian was appointed Chief of Staff to the Inspector of Motorized Units, Major-General Lutz—a fellow tank enthusiast. By now the former had a clear idea of the type of tanks he wanted to smash their way through the defences, or outflank them, and exploit strategically. He envisaged a light and medium model, both to be equipped with machine-guns, an armour-piercing gun and radios, and to be capable of a speed of about 25 m.p.h. These would eventually begin to appear in 1938 and 1939 as the Panzer Mark III and IV respectively. In the meantime the Panzer I was ordered for training purposes in 1932 and began to appear in 1934. This was a two-man, five-ton mini-tank which followed the design of the British Vickers Carden Loyd light tanks. It was armed only with machine-guns and could travel at a maximum speed of 24 m.p.h. Its attractions were speed, reliability and cheapness but it had little fighting power and was a dubious investment. Some 1,500 tanks of this type were produced before the outbreak of war. Another stop-gap, the Panzer II, was ordered in 1934. It weighed nearly nine tons, had a crew of three and was armed with a 20mm gun as well as a machine-gun; this was a surprisingly weak armament, since comparable light tanks in other countries were being equipped with a 37mm gun. What is even more surprising in retrospect is that these models, under-gunned and thinly protected, still provided the great majority of German tanks in the Polish campaign in 1939.[22]

In the autumn of 1934 the long-awaited trebling of the German army began: twenty-one infantry divisions were formed and the first tentative steps were taken towards the creation of an armoured spearhead with the raising of the First Panzer Brigade. This consisted of three regiments with two battalions each, equipped with Panzer Is as they became available. When Lutz was promoted to the Command of Panzer troops in June 1934 Guderian once again went with him as Chief of Staff. It is likely that the latter profited from Liddell Hart's writings by studying his articles describing Hobart's experiments in deep penetration with the First British Tank Brigade in 1934 and 1935.[23] The British tank exercises provided important ideas for the first German manoeuvres with a complete Panzer division, which were held at Munster Lager in July 1935. The manoeuvres, which were attended by the War Minister, Blomberg, and the Commander-in-Chief, Fritsch, were deliberately calculated to banish the still considerable opposition to the idea of Panzer divisions and corps. As Guderian explained:

> It was not our intention on this occasion to instruct the subordinate unit commanders in the appreciation of and reaction to their individual tactical problems, but rather simply to demonstrate that the movement and commitment in action of large masses of tanks together with supporting weapons, was in fact possible.

The exercises were a great success, so much so that at the conclusion Fritsch remarked jokingly: 'There's only one thing missing. The balloon [signalling the end of the exercise] should have *Guderian's Panzers are best* marked on it.'[24] By this time Guderian was passing beyond the theoretical stage when Liddell Hart's ideas had been useful to him, and henceforth he was probably guided largely by his practical experience in the field.

The year 1935 was marked by four significant events relating to the development of German armoured forces. Firstly, Hitler denounced the Versailles Treaty, thus removing the last vestige of the restrictions on the German production of tanks. Secondly, in the summer the first Panzer division was improvised from existing units. Next came the establishment of *Panzertruppen* as a distinct arm and, finally, in October the formation of the first three permanent Panzer divisions, with Guderian in command of the Second. It should not be imagined, however, that all was plain sailing after this. Only a small fraction of the supporting units could be mounted on unarmoured tractors, so that their ability to follow the tanks across country was very restricted. This far from ideal situation was never completely remedied throughout the war. 'Even the best-equipped divisions,' O'Neill writes, 'only reached a stage where a whole Panzer grenadier regiment (i.e. half of the brigade), some anti-aircraft artillery and some field artillery and parts of the signals and engineer support units were on half-tracks.'

Moreover, as Guderian explains, there was still a long, hard battle to be fought before the German High Command, under pressure from Hitler, finally took the decision to commit all the tanks to the role of deep penetration. In 1936 and 1937, largely under the influence of the conservative Chief of Staff, Beck, the Panzer brigades were assigned to the role of infantry support;

while the cavalry's demand for greater control of motorized troops was acceded to by the creation of three Light divisions instead of three additional Panzer divisions. In 1938, however, partly as a result of Hitler's direct interference in control of the armed forces after the removal of Blomberg and Fritsch, there was a resurgence of the *blitzkrieg* doctrine. In February Guderian was appointed to the command of XVI Corps (containing three Panzer divisions), and in the following November, at Hitler's own suggestion, he was made the first Commander of Mobile Troops with privileged right of direct access to the Führer whenever he felt himself obstructed on a vital issue.

By September 1939 six Panzer Divisions were in being and Guderian had been primarily responsible for training them in the techniques of *blitzkrieg* by which Poland was overwhelmed. Although the Polish tank forces were completely outnumbered and outclassed, the significance of the campaign should not be underestimated. It represented a major effort for the German Army and the first battle test of the Panzer divisions. The results proved the soundness of Guderian's ideas and established the Panzer divisions as the mobile spearhead of the ground forces capable of smashing through strong defences *as well as* rapid outflanking manoeuvres and deep penetration into the enemy's rear areas.[25]

After the Second World War Liddell Hart became personally acquainted with many of the senior German commanders, such as Field Marshals Rundstedt, Manstein and Brauchitsch, who were prisoners of war in England, and he later met others, such as Guderian, who were imprisoned on the Continent. Many of them paid tribute to his pre-war and wartime influence on the German army in their correspondence and memoirs, and on the signed photographs which they sent him, sometimes at his request. Without presuming to challenge the sincerity of any particular tribute, it may be suggested that collectively they must be treated with caution, since they were made after the event and in circumstances where objectivity was difficult. As mentioned in a previous chapter, Liddell Hart made himself a leading champion of the captured German generals in the five years or so after the ending of the Second World War in the sense that he campaigned in the Press for better prison conditions and against their being tried as war criminals. In private he behaved with great generosity towards them, sending them cigarettes, tobacco, food parcels and copies of his books and articles. In particular hard cases he put himself to even further trouble, supplying the ailing Rundstedt with a mattress and securing the transfer of Manstein's wife and son to the home of the Field Marshal's sister in the French-occupied zone. These actions were clearly motivated chiefly by feelings of compassion for a beaten enemy who, in his opinion, had fought honourably and with great skill. At the same time, however, he was intensely interested, both as a historian and a participant in events, to glean every scrap of information possible about their pre-war ideas and conduct of operations in the Second World War. The success of his indefatigable enquiries is evident in a whole cabinet and shelves packed with correspondence and interview notes that formed the basis of his book *The Other Side of the Hill* (1948).

No one who knew him could deny that Liddell Hart was rather naïvely vulnerable to flattery, and whether or not any individual German generals exploited this foible deliberately is beside the point: what matters is that their relationships with him were coloured by warm feelings of gratitude and, in some cases, mutual admiration. Robert O'Neill expressed the point well in his obituary article:

> They [the German generals] were only too keenly aware of their own relative friendlessness in the atmosphere of defeat and they assiduously cultivated Liddell Hart's acquaintance. He, of course, was gratified to meet and talk with military leaders who had proved the success of his own ideas and who, in themselves, represented a valuable historical source.[26]

The implication behind the second sentence quoted above becomes clearer when it is recalled, as explained earlier in this book, that Liddell Hart's reputation had been in eclipse during the Second World War, and in the later 1940s he was still feeling extremely bitter about his neglect by the British military and political establishments. This is not to imply any cynical intent on Liddell Hart's part in deliberately trying to boost his reputation; simply that he was gratified to receive confirmation that his prophetic military ideas *had* been fully appreciated on 'the other side of the hill'.

Incidentally, although he was later to declare *à propos* of Israeli tributes that he had never felt wholly comfortable when praised by the Germans, this unease is not apparent in the 1940s. On the contrary, he was very proud of the German tributes and extremely sensitive to the slightest implication that his influence in that quarter was open to doubt. Thus when Colonel (now Field Marshal Sir Michael) Carver incautiously used the phrase 'attempt to prove' (of his influence on the German generals) in reviewing *The Rommel Papers*, Liddell Hart showed himself deeply hurt and persuaded Colonel Carver to modify his statement along the lines suggested by Liddell Hart in a letter to the editor of the *Royal Armoured Corps Journal.*

> For my part [Liddell Hart wrote to Colonel Carver] I have repeatedly emphasized that the development of such forces and their techniques was not a solo effort, but a combined one—and that Guderian's tribute should rightly be shared by the whole band of crusaders in or associated with the Royal Tank Corps.[27]

Liddell Hart did not correspond with Guderian until September 1948, when he sent him a copy of *The Other Side of the Hill*, and they did not meet until Liddell Hart visited Germany in 1950. It was through his meetings with General von Thoma, who had commanded the German tank units in Spain during the Civil War and had been taken prisoner while in temporary command at El Alamein, that Liddell Hart learned that Guderian had been a keen student of his publications. After an interview with von Thoma at Grizedale prisoner-of-war camp in November 1945 Liddell Hart noted:

> The German tank officers keenly studied British military writings on armoured warfare, particularly my own. They also followed with keen interest the pioneer activities of the original Tank Brigade in the British Army ... permanently formed in 1934 under Hobart.

Asked in 1949 when he had first read Liddell Hart, Guderian could only recall 'I think that I first read your articles about the year 1923–24. I read *When Britain Goes to War*, *The Future of Infantry* and *The Remaking of Modern Armies*'.

It was also von Thoma who helped Liddell Hart to establish the point, mentioned earlier, that Guderian had used reports of the British tank manoeuvres of 1934 and 1935 as a blueprint for the training of his own Panzer division. In October 1941 Liddell Hart received a letter from a Mrs Zina Hugo, the wife of a British Tank Corps officer. In August 1939 Mrs Hugo had been staying at the Bulgarian officers' club in Pleven, where she met a Colonel Khandyeff who expressed a keen interest in Liddell Hart's publications on tank tactics and in Colonel Hobart's tank manoeuvres. Mrs Hugo related Colonel Khandyeff's experience on attachment with a German armoured division a few years earlier as follows:

> The divisional commander was absolutely mad about the exploited and unexploited possibilities of tanks. His faith in armoured formations was such that he took a tremendous amount of pains in planting the same enthusiasm in the people under him.
>
> He spent his own money on providing copies of foreign books and periodicals, as well as on the services of a local tutor for the rough translations. His gods were General Fuller and Captain Liddell Hart. Liddell Hart, he considered, was the best analytical brain in the world, and his articles translated, read, and studied, were discussed long before they would be vetted and sent from Berlin.
>
> While with him, Liddell Hart's accounts of the manoeuvres began to appear in *The Times*. As much as possible, every move of the manoeuvres was copied and put into practical demonstration. It was like the rehearsal of a play. The General was the happiest and busiest man, saying that Hobart gave him an answer to so many queries—and an inspiration. When a visiting anti-tank expert spoke of tank limitations as well as tank or no-tank country, quoting various opinions, including those of well-known people in England, the General impatiently dismissed him by saying—'It is the old school, and already old history. I put my faith in Hobart, in the new man'.[28]

Rather surprisingly in view of the circumstantial detail, either Colonel Khandyeff omitted the German general's name or Mrs Hugo did not remember it; the latter seems more likely, since Guderian's name would not have meant anything to an English lady in August 1939. Von Thoma, however, said that he remembered Khandyeff's visit well and thought the general referred to was probably Guderian. In 1948 Liddell Hart sent Guderian a copy of Khandyeff's account as related by Mrs Hugo and asked for his comments. Guderian replied:

> I don't remember his being present at the manoeuvres of my Panzer division. But as Thoma remembered meeting him and as I was Thoma's divisional commander at the time I think that Khandyeff was speaking about me in his letter.

A few months earlier Guderian had written to General Dittmar (with whom Liddell Hart was in contact) that he could not swear it was his division Colonel Khandyeff referred to:

> It is certain however that I had read many articles by Liddell Hart and that they were always of burning interest to me [dass sie mich stets glühend interessierten] and that I have learnt much from them.[29]

Liddell Hart never met or corresponded with Field Marshal Rommel (who, it will be remembered, was forced to commit suicide in the autumn of 1944 as a result of being implicated in the July 20 Plot against Hitler). The evidence that he was Liddell Hart's 'disciple' is therefore necessarily more tenuous and was derived initially from General Fritz Bayerlein, Operational Chief to Guderian's Panzer Group in 1940 and Rommel's Chief of Staff during part of the North Africa campaign. When contacted via Rommel's son Manfred, Bayerlein informed Liddell Hart in 1950:

> During the war in many conferences and personal speeches [*sic*] with the late Field Marshal Rommel we discussed your military works that gained our admiration. We recognised you as a military author who made the greatest impression on the Field Marshal and who highly influenced his tactical and strategical conceptions. As the former Chief of Staff to Rommel I can state not only General Guderian but Rommel too could be called your pupil in many respects. Your books: *The Future of Infantry*, *Dynamic Defence*, *When Britain Goes to War* [and] *Europe in Arms* gained our recognition especially whilst your standard work *The Strategy of Indirect Approach* was still unknown to us at that time.

Bayerlein added a similar note in *The Rommel Papers* (1953), which he helped Liddell Hart to edit. Rommel's only specific mention of Liddell Hart refers to a wartime article which he read in 1942. Manfred Rommel recalled that:

> My father had a great admiration for you and I remember that he

> read one or more of your books when he was appointed commander of the 7th Armoured Division in France [in 1940]. In North Africa he studied the articles you wrote during the war. I was too young and too much of a layman to keep in mind my father's remarks on your books.

This evidence seems reasonably conclusive on the point that, unlike Guderian, Rommel was not a follower of Liddell Hart and the mechanized warfare school before 1939 (von Thoma went so far as to call him an 'opponent' of their concepts and added that he only understood the tactics and not the techniques of tanks) but that he *did* profit from their publications in and after 1940.[30]

Although Liddell Hart never numbered Field Marshal Erich von Manstein among his disciples, he did attempt to establish a particular claim to influence which is not substantiated by his Papers. This concerned Manstein's all-important idea in the winter of 1939 that the Ardennes were not impassable for large mechanized forces and that decisive results could be obtained against France by using that route to achieve surprise.

Liddell Hart's account of his prediction regarding the viability of the Ardennes may be summarized as follows. After a tour of the area in 1928 he came to the conclusion that the *Allies* had been mistaken in the First World War in believing the Ardennes to be completely unsuitable for a counter-offensive. He referred to this deduction in his books *The Decisive Wars of History*, *The Real War* and *Foch*. In the first-mentioned of these books, for example, he referred to the Allied plan of September 1918 designed to cut off a large part of the German armies in France by a wide pincer movement. This hope, he added,

> was based on the idea that the Ardennes formed an almost impassable back wall with narrow exits on the flank. One may add, incidentally, that this idea of the Ardennes must have arisen from a lack of knowledge of the district, for it is well-roaded, and most of it is rolling rather than mountainous country.

In the mid-1930s he was still thinking in terms of a possible *French* offensive and considered that the German plan 'was more likely to be one of luring the French into a trap than of taking the offensive'. In a private discussion with Brigadier Sir Ronald

Adam and Colonel Bernard Paget on 12 May 1936 he suggested the Germans might *counter-attack* through the Ardennes if the French advanced through Belgium, but of course this was unknown to Manstein. Finally, in *The Defence of Britain* (1939) he made veiled references to methods by which the Belgians and French might block the narrow defiles through the mountains by felling trees and carrying out demolitions.

> Much of the Ardennes [he wrote] is akin to the contours of Salisbury Plain, though more heavily wooded, but the rivers cut much deeper furrows, and at many points where the roads cross them a handful of machine-guns might hold up an army corps.

In this survey of the French and Belgian frontiers Liddell Hart stressed the advantages of the defenders. The Ardennes, he remarked, might prove a strategic trap for an invader if he failed to cross the Meuse. He also noted that Belgium was most vulnerable to a surprise attack on her Dutch flank, and if the main German advance in 1940 had come through the Netherlands he could have cited this passage as an accurate prophecy.[31]

It is perfectly possible that the Germans came across these references and interpreted them from their own standpoint, but no evidence has been found to suggest that they did so. After all, many of them, including Guderian, knew the area personally from the First World War and could have drawn their own conclusions. It is perhaps significant that in 1948 Manstein recounted to Liddell Hart in some detail how he had conceived the plan and got it accepted by Hitler without mentioning Liddell Hart's references to the Ardennes. At the end of the Manstein file there is a note referring to Reginald (now Lord) Paget's book *Manstein: His Campaigns and His Trial* in which Liddell Hart implies that Manstein had acknowledged his influence over the idea of the Ardennes route in conversation with Paget but later retracted the admission:

> Paget subsequently told me that Manstein raised objection to such a definite statement of what he had told Paget privately—so in the published version the sentence was revised to read 'Captain Liddell Hart, he said, had suggested in an article before the war, that an armoured thrust through the Ardennes was technically possible'.[32]

Two further tributes may be cited as typical of the praise which Liddell Hart received from former German commanders in the late 1940s and 1950s. General F. W. von Mellenthin, another former Chief of Staff to Rommel and author of *Panzer Battles* (1955), was quoted as follows by the *Evening Standard* on a talk he gave at the Safari Club in 1959:

> 'It may interest you to know', said the General, with only the faintest echo of a laugh up his sleeve [*sic*], 'that we mostly studied the books of English tank experts like Captain Liddell Hart. German tank successes in the first year of the war were mainly due to the fact that we adopted the theories taught by Captain Liddell Hart while the enemy, in their thoughts, were still in the First World War'.[33]

General Blumentritt, Deputy Chief of the General Staff in 1942 and Chief of Staff in the West 1943–44, thus summarized the influence of the British school of armoured theorists in pre-war Germany in an article published in 1949:

> Liddell Hart and Fuller were for us young officers after 1920 'the modern military authors'. Particularly in the Reichswehr they were carefully studied and all their articles read. In those days we were lieutenants and captains aged 28 to 35 and took delight in the modern spirit of these writers. At the same time we were irritated by the negative attitude of our senior generals, whose mentality and way of thinking circled round the old traditions of the horse. They looked down upon technics as unsoldierly craftmanship. But we knew, even then, that authoritative circles in England too were against the modern conceptions. We, therefore, knew differences of opinion. In our camp only Guderian and von Reichenau fought for the mechanization of warfare. The other high generals before 1939, and the bulk of the General Staff, faced these revolutionary developments with scepticism, including the then Chief of General Staff, Beck. It was a struggle between the antiquated, aristocratic soldier of the 19th century and the youthful technician soldier of our age.[34]

At a time when he was feeling particularly frustrated after the Second World War, Liddell Hart understandably exaggerated the contrast between the neglect of his ideas by British generals and their enthusiastic reception by the Germans. In fact he had his supporters in the British Army, while his influence in Germany was largely due to three pro-Nazi generals, Blomberg, Reichenau and Guderian, who were all well placed in the 1930s to promulgate his ideas.

Three points of particular importance emerge from this chapter. The first is that Fuller deserves at least as much recognition as Liddell Hart for propagating the gospel of mechanization in Germany in the 1920s and early 1930s. Fuller's name was particularly associated with tanks, whereas Liddell Hart was highly regarded as a military historian and commentator on defence issues generally. Secondly, Liddell Hart's most valuable and distinctive contributions to German military thought were probably made *before* the creation of the first Panzer divisions; specifically in his concepts of deep strategic penetration by massed armoured forces, and his insistence on the need for an infantry component and mechanized supporting arms as distinct from all-tank divisions. Finally, Liddell Hart's post-war claims to have exerted an important influence on the German army were essentially due to the ackowledgements of the man who chiefly inspired the creation of Panzer forces in the 1930s and then played the leading role in their most brilliant campaign in France in May 1940, General Heinz Guderian. It is fitting to conclude with Guderian's generous tribute to Liddell Hart in the encyclopaedia *Der Grosse Brockhaus*: 'The creator of the theory of the conduct of mechanized war.'[35]

Notes

1. A. Rapoport (ed.) *Clausewitz on War* (Pelican Classics, Penguin 1968) p. 191.

2. 9/24/61 General Leo Geyr von Schweppenberg. Geyr to Liddell Hart 10 August 1935, 14 February, 16 and 27 May 1936.

3. Richard Ogorkiewicz and Robert O'Neill made this point independently in letters to the author on 31 January and 6 February 1976 respectively. In my account of Guderian's use of Liddell Hart's ideas I have closely followed Robert O'Neill's essay 'Doctrine and Training in the German Army, 1919–1939' in Michael Howard (ed.) *The Theory and Practice of War* (Cassell 1965: Praeger 1966) pp. 145–165.

4. Ogorkiewicz and O'Neill both stress this point in the letters cited above.

5. R. Ogorkiewicz *Armour* (Stevens: Praeger 1960) p. 264.

6. Ibid. pp. 206–207. Liddell Hart *The Tanks* I (1959) pp. 165–168. R. M. Ogorkiewicz 'Marginal Notes: German Tank Development and General Guderian's "Panzer Leader"' in *Royal Armoured Corps Journal* October 1952 pp. 196–206.

7. 9/24/124 Manstein to Liddell Hart 25 January 1958.

8. 11/1932/9 talk with General Blomberg at Geneva 8th March 1932. See also the numerous index references to Blomberg and Reichenau in *Memoirs* I.

9. 9/24/87 Reichenau to Liddell Hart (holograph) 28 November 1932. General Sir Andrew Thorne to Liddell Hart 13 October 1942. O'Neill op. cit. p. 154.

10. 11/1931/28.

11. Fuller to Liddell Hart 28 February 1935.

12. Sir Maurice Hankey to Liddell Hart 27 December 1933. 11/1948/38 Thorne to Hankey, n.d. but probably 1946 when Liddell Hart was gathering testimonies to support his candidature for the Chichele Chair in the History of War at Oxford; see Liddell Hart to Thorne 18 March 1946.

13. Liddell Hart to Thorne 18 July 1942.

14. Letters to the author from Klaus Gütig 24 September and Dr Robert O'Neill 15 December 1975.

15. 9/24/105 von Geyr to Liddell Hart 8 March 1949. 9/24/125 Manteuffel to Liddell Hart 10 March 1949.

16. H. Guderian *Panzer Leader* (Joseph: Dutton 1952) pp. 19–46; O'Neill op. cit. Ogorkiewicz 'Marginal Notes' op. cit. corrects some points in Guderian's account of the technical development of German tanks.

17. *Panzer Leader* p. 20.

18. O'Neill pp. 152–153.

19. 11/1948/34 'Early Ideas on the Pattern of the Armoured Division'. Ogorkiewicz 'Marginal Notes' op. cit. p. 198. Fuller is usually associated with the 'all-tank' school but at times he too stressed the importance of close co-operation with infantry; see his comments on the 1936 German manoeuvres in *Machine Warfare* (Hutchinson 1942: Infantry Journal 1943) p. 55.

20. Ogorkiewicz *Armour* pp. 72–73. O'Neill pp. 156–159.

21. The source for Beck's remark was Peter Paret, whose father was acquainted with Beck's friends Professors Meinecke and Sauerbruck; see 'Historical Note' 1 October 1958, 13/5 miscellaneous.

22. Ogorkiewicz *Armour* p. 212. On 1 September 1939 the German armoured forces comprised six regular Panzer divisions, one improvised division and four light divisions. On that day the total number of German tanks (excluding Czech vehicles) amounted to 3,195: 1,445 Panzer Is, 1,226 Panzer IIs, 98 Panzer IIIs, 211 Panzer IVs and 215 command tanks.

23. For Liddell Hart's belief that his account of British tank exercises had influenced Guderian see *The Tanks* I pp. 317–321, 347–348.

24. *Panzer Leader* pp. 35–36.

25. Ibid. p. 37; O'Neill p. 158; Ogorkiewicz *Armour* pp. 73–75, 211.

26. O'Neill 'Sir Basil Liddell Hart: An Appreciation' in *Army Journal* (Australian) April 1970 pp. 29–39.

27. Colonel Carver to Liddell Hart 8 and 24 December and Liddell Hart to Carver 15 December 1953.

28. 9/24/144 talk with von Thoma 1 November 1945. 9/24/62 Liddell Hart to Guderian 3 March 1949 and Guderian to Liddell Hart 19 March 1948. Zina Hugo to Liddell Hart 28 October 1941 (copy) 13/5 miscellaneous file. See also 11/1948/38, which includes an extract from Khandyeff's account of his attachment to a German division and which Liddell Hart cited in support of his claim to have influenced Guderian.

29. 9/24/62 Liddell Hart to Guderian 20 and 28 September 1948. Guderian to Liddell Hart 7 October 1948. Extract from a letter from Guderian to Dittmar (in German) 29 August 1948.

30. 9/24/50 Bayerlein to Liddell Hart 15 February 1950. 9/24/24 Manfred Rommel to Liddell Hart 28 December 1949. 9/24/144 talk with von Thoma 20 November 1945. Liddell Hart (ed.) *The Rommel Papers* (1953) pp. 203, 299, 520.

31. 11/1948/28 'The Ardennes as a Potential Route for Mechanized Forces'. 11/1934/1 Diary notes 13 April. *Decisive Wars of History* p. 225. *The Defence of Britain* pp. 216–219.

32. 11/1948/32 'Attack and Defence' n.d. 'Manstein, who conceived the plan of the tank stroke through the Ardennes, has stated that the idea came to him from an article of mine.' 9/24/71 Manstein to Liddell Hart 24 February 1948. 9/24/62 Guderian to Liddell Hart 2 January 1949. 9/24/124 Liddell Hart to H. A. Jacobsen 29 February 1956. The last undated item in this file contains the reference to Manstein and R. T. Paget. See also the latter's *Manstein* (Collins 1951) p. 22. Lord Paget believes that in his first draft he stated that the idea for the Ardennes offensive originated with Liddell Hart and that Manstein queried this point in correcting the text: letter to the author 21 January 1976.

33. 9/24/73 cutting from the *Evening Standard* 27 August 1959. See also F. W. von Mellenthin *Panzer Battles* (Cassell 1955) p. xvi. Mellenthin's tribute is quoted as a typical example of German generals' references to Liddell Hart after 1945, but he is not a particularly reliable witness.

34. 11/1948/38. For a similar tribute see S. Westphal *The German Army in the West* (Cassell 1951) p. 37.

35. Quoted by O'Neill (and his translation) p. 164. Cf. the literal translation in 11/1948/38 'British military writer, described by Guderian as the theoretical originator of mechanized warfare'.

9

Liddell Hart's Influence on Israeli Military Theory and Practice

There can be no doubt that Liddell Hart's name and the main military ideas associated with him are still familiar to a wide variety of Israelis. Indeed, in comparison with the reputation of foreign military pundits in Britain, he is astonishingly well known, as the author was able to check for himself in September 1975. But it must be remembered that the vast majority of adult Israelis are serving, or have served, in the armed forces, and that for obvious reasons war and politics are discussed there with as much interest as football and cricket in certain more secure countries. Also, on closer examination, one finds that in most cases knowledge of Liddell Hart has been acquired second-hand from newspapers and journals; in particular, his official visit in 1960 and his comments on the 1967 war both received great publicity. Only two of his books, *Scipio* and *Strategy: the Indirect Approach*, and a handful of his articles have been translated into Hebrew.[1] Nevertheless, despite these reservations, there are good grounds for believing that, in general terms, Liddell Hart's ideas have made a significant impact in Israel—the most important evidence being the tributes of several distinguished soldiers and scholars. What needs more careful examination, however, are the sources, extent and nature of this influence. It is simply not good enough, as some enthusiastic but historically unsophisticated interviewers and journalists have done, to attribute all the Israelis' successes to their skilful employment of the indirect approach as inculcated by Liddell Hart.

If the Israelis have indeed been Liddell Hart's 'best pupils', this has to a large extent been predetermined by their unique circumstances. The state of Israel was literally born in war and has survived precariously ever since, surrounded by enemies who would like to see it destroyed. Morale has been helped mightily by the knowledge that one major defeat would almost certainly entail national extinction. Israeli strategy has also been decisively influenced by geography: until the victories of

1967 her frontiers were militarily nonsensical and certainly allowed no option of defence in depth and reliance on a counter-offensive. In particular, with a 'waist' tapering to scarcely ten miles north of Tel Aviv, some of the chief population centres were within enemy artillery range and the country was in danger of being cut in two by a sudden attack from the east. Even with the territory gained in 1967 Israel's cities are still within close flying range of enemy airfields. Before the 1967 war, for example, Tel Aviv was about twelve minutes' flying time from El Arish, and it is still only half an hour from Cairo. Moreover Israel's economy is extremely fragile. Despite determined efforts to create an industrial base with a capacity to manufacture armaments, she is still critically dependent on imports in the defence field, especially aircraft and tanks. In sum, Israel can contemplate only very short wars (the War of Independence 1948–49 was intermittent and of low intensity); even victory in a war lasting more than a few weeks would be economically disastrous. As a corollary to the last point, Israeli strategic planners have had to assume that the Great Powers would not let a long war of attrition develop. In theory this might of course save Israel from annihilation, though so far in practice it has prevented her from pressing home her victories. But the message has clearly been that it is vital to gain military advantages within the first few days as diplomatic bargaining counters.

Lastly, and perhaps most important of all, the Israelis have a uniquely low tolerance of military casualties. Their utterly unmilitaristic attitude in this respect is probably a compound of three factors: the Jewish history of persecution, the small size of the population, and the closely-knit social structure of the extended family. Even after nearly thirty years of constant readiness for war, the Prime Minister is still informed immediately of every casualty, and photographs with a brief biography of every dead soldier are published on the front pages of the newspapers. Indeed this admirably humane attitude has been carried so far that it could constitute a military handicap. The Cabinet and the General Staff may be inhibited from opting for bold military operations where there is a risk of higher casualties. Certainly there are numerous examples of strong press criticism even over highly successful raids where the losses were considered excessive.[2] Though the outcome of the 1973 operations was a triumph in military terms, the comparatively high losses—by Israeli standards—of about 3,000 dead and 8,000 wounded have caused bitter criticism and recrimination. On the other hand the knowledge that every individual's life is considered precious by the military leaders and that every care will be taken to minimize casualties and save the wounded gives a tremendous boost to morale. Liddell Hart commented on the paradox after the 1967 war that although the Israeli people were unwilling to tolerate high casualties, her soldiers' greater willingness to die for their cause gave them a tremendous advantage over their opponents.[3]

These diverse factors in combination have clearly exerted a decisive influence on the makers of Israeli strategy. In simple terms, the Israeli forces must opt for a quick and decisive style of operations while practising a ruthless economy in both manpower and *matériel*. Circumstances have forced them to rely primarily on quality rather than quantity, particularly as regards personnel. Thus the emphasis in 1956 and 1967 (also recommended by the military advisers in 1973) has been on a sudden, unexpected first strike designed to catch the enemy off-balance and gain a decisive advantage in the first hours of a conflict. Liddell Hart's writing can certainly be cited in support of the surprise element in this doctrine, but on the whole he tended to argue that a first strike seldom wins the war and that the ideal strategy is the *counter-offensive* after the attacker's impetus had been checked, such as Joffre's riposte on the Marne in 1914. The Israelis implemented this strategy very successfully in 1973, but not from choice.

There is another broad consideration which must be mentioned before examining Liddell Hart's influence in detail. From rather unpromising military origins, the Jewish settlers in Palestine, and later immigrants from Europe and North Africa, gradually developed military skills of a high order.[4] After the creation of the state of Israel had supplied a permanent and legal structure, an army emerged which was almost ideal in terms of the criteria which Liddell Hart had been proclaiming in the far less favourable European context since the 1920s. Here, in short, was a small democratic country unimpeded by the weight of a regular military tradition, with a citizenry above average in idealism and education and the value it attached to intelligence, possessing a fierce military spirit yet remarkably unmilitaristic in the pejorative sense, and so determined to survive that it was prepared to be constantly self-critical even though victorious.[5]

The lack of a Jewish military tradition since the Biblical era, as distinct from the valour displayed by numerous Jews in the armed forces of the other states, actually proved to be an advantage in Israel in that it allowed scope for original methods and new ideas. In the egalitarian social atmosphere that prevailed, especially in the kibbutz, orders could rarely be imposed by the simple authority of military rank; open debate was accepted as natural. Similarly, the diverse origins of the Jewish fighting forces before 1948 militated against the imposition of a uniform system of command. A decentralized command system was retained in the permanent defence forces after 1948, providing scope for flexibility and individuality. Not least important, a good deal of the Palmach tradition[6] was retained; instead of breaking civilians' individuality in order to turn them into soldiers, Palmach training encouraged initiative as exemplified by the slogan 'The smallest unit is the single man with his rifle'. It also stressed that officers should lead by personal example from the front rather than by detailed orders; hence the catchphrase 'follow me' which is still popular.

As regards military service no country asks more from its citizens than Israel. Conscription varying from two to three years, followed by frequent periods of reserve training and refresher courses, is essential because reservists have to be prepared to serve in front-line formations with only a few hours' notice. Similarly, reserve officers have to acquire the skills of full-time career officers in other armies. Where but in the Israeli Defence Forces would you find a divisional commander drawn straight from the reserve?[7] Yigael Yadin, Chief of Staff 1949–52 and subsequently an internationally renowned archaeologist, is said to have coined the phrase: 'Every citizen is a soldier on eleven months' annual leave.'[8]

The Israeli Defence Forces have achieved a remarkable state of preparedness and efficiency with a minimum amount of militarization or segregation between officers, N.C.O.s and other ranks. Officers' uniforms are not markedly different from those of the men, and the idea of officers' clubs did not catch on. 'In the reserve formations, the atmosphere remained resolutely civilian in the midst of all the trappings of military life . . . Cases of university professors who serve under the command of their students, or clerks who carry higher ranks than their bosses, are common-place and so no longer attract attention. Reserve officers are proud of their rank status *vis-à-vis* their families and friends, but among their fellow reservists it usually gives them little or no status advantage.'[9]

Liddell Hart, though a vehement opponent of conscription, recognized that in Israel its effect was to 'civilise the military'. Here was a 'humanist army' in which citizens retained their civilian values while attaining a high degree of professional skill. After the 1967 war he quoted at considerable length a letter from an Israeli correspondent who remarked that whereas in America many of his friends thought of the Army 'as a last resort for those too stupid to avoid it, in Israel people who are rejected by the Army sometimes leave the country—so great is the disgrace'.[10]

Liddell Hart made no pretence of impartiality where Israeli military affairs were concerned; indeed his attitude rather resembled that of a proud headmaster praising successive generations of pupils who not only carried off the prizes but improved upon his own teaching. He was at pains to emphasize that his pro-Zionist sentiments pre-dated the foundation of the state of Israel. In 1937 and 1938 he had been consulted in London by a number of Zionist leaders about the strategic aspects of the Middle East situation. Among them were Chaim Weizmann, the future President of Israel, David Ben Gurion, Israel Galili and Moshe Sharett (later Foreign Minister and Prime Minister). In his *Memoirs* he recorded that 'from the time I first met the budding leaders of Israel I found there a grasp of military problems and new military ideas comparable to that of the Germans, and in some respects surpassing theirs'. Two or three years before this, Generals Dill and Wavell, successively Commanders-in-Chief Palestine, had passed on to Liddell Hart their very high opinions of the fighting abilities of the Palestinian Jews. Towards the end of 1938 Liddell Hart also had two meetings with Captain Orde Wingate to discuss the tactics and training of the counter-guerrilla units (the famous 'Special

Night Squads') which he had organized that year in Palestine from Jewish volunteers and British regulars to deal with the Arab armed bands which had been causing the British garrison so much trouble since 1936. According to Liddell Hart's *Memoirs*, Wingate had been using his early works on infantry training and also his more recent booklet *The Future of Infantry*.[11] Thus in this important respect Liddell Hart does not appear to have been swayed by the pro-Arab sympathies of his hero T. E. Lawrence, to whom incidentally Wingate was proud to claim a distant kinship.[12] One can only speculate that here, as on so many issues, Liddell Hart was reacting against the conventional outlook of the British military establishment, as Wingate certainly did in his fanatic espousal of the Zionist cause.

In an attempt to be as precise as possible in evaluating Liddell Hart's influence on the Israelis I have relied first and foremost on the files of those who corresponded with him and have checked the information gleaned from this source by personal interviews. Since all but two of the important correspondents, Jaacov Dori and Israel Beer, are still alive, this was a rare opportunity to secure first-hand evidence of the kind of influence that the pen can exert on the sword. It should perhaps be underlined that what follows is in no sense an attempt to delineate the military or social history of the Israeli Defence Forces. A working knowledge of these must be assumed so that a minimum of information will suffice to provide a framework in which Liddell Hart's ideas can be discussed.

Despite the tacit, and in many quarters reluctant, co-operation of the Jewish Defence Force (the Haganah) with the British in the form of the Special Night Squads and (Jewish) Settlement Police, it remained an illegal, though tolerated underground army. By the outbreak of the Second World War the Haganah had acquired a rudimentary organization of paid staff officers and a Chief of Staff in Jaacov Dori. But it still lacked such vital ingredients of a regular army as an adequate supply of weapons, men trained for sustained action and a cadre of experienced officers. Training could only be carried out by small groups of guerrillas based on the kibbutz. Britain, opting in May 1939 for a policy which would give the Arabs a decisive advantage in an eventual Arab–Jewish state, rejected the Zionist proposal to form a Jewish army or even a division, but as many as 27,000 Jews volunteered to serve in the British and Allied forces and a Jewish Brigade formed part of the British Eighth Army in Italy in 1944–45. When Rommel's offensive reached the gates of Egypt in 1941 the new leaders of the Haganah decided to raise a mobile force of full-time soldiers for the defence of Jewish territory, known as the Palmach. This élite force, under the command of the veteran Yitzhak Sadeh, became the future training ground for many of the future leaders of the Israeli army, including Yigal Allon and Moshe Dayan. Though strictly under Haganah command, the first two companies took part in the British invasion of Syria and Lebanon in August 1941, but this was virtually the last instance of Anglo-Jewish co-operation in the Middle East. The Palmach was élitist and openly left-wing in ideology; it was essentially a youthful force of guerrilla infantry which achieved high standards in individual skill at arms and fostered group morale. For the remainder of the war the Jewish political leadership which controlled the Haganah was in an ambiguous position—increasingly at odds with the British in its determination to establish a viable Jewish National Home but willing in the short term to co-operate to defeat the Axis. It could not however control small groups of Jewish extremists such as Lehi (Lohamel Herut Israel—Hebrew for Israeli Freedom Fighters; known to its detractors as the Stern Gang) or the I.Z.L. (Irgun Zva'i Le'umi—Hebrew for National Military Organization), which were frankly terrorist organizations.[13]

It was in this confused period before the establishment of the state of Israel that the seeds of Liddell Hart's influence were sown. They were sown by one of the most brilliant of Israel's soldiers and one of the very few who, before 1948, was fluent in English—Yigael Yadin. A member of the political élite (Yishuv), he was born in 1917, the son of a distinguished professor of archaeology. In the late 1930s he acquired practical experience in command of the Jerusalem area of the Fosh or paramilitary Jewish force raised to assist the British police. In the War of Independence he was nominally Head of Operations in the General Staff Branch but for all practical purposes was Chief of Staff, a post which he was to hold officially from 1949 to 1952, when he left the Army to follow in his father's footsteps and achieve international fame as the excavator of Masada.

Apart from the tactical ideas which had been studied by Wingate, it was Yadin who played the crucial role in introducing Liddell Hart to many of the future military leaders of Israel.[14] From about 1940 to 1943 Captain Yadin was in charge of the planning branch of the Haganah secret officers' school which was situated on a kibbutz near Mount Carmel. Among those who took the two-month course were many young men who later achieved distinction in the I.D.F., including at least two Chiefs of Staff in Zvi Zur and Yitzhak Rabin.

Yadin, who read French and German as well as English, studied voraciously to find suitable material for his course and in this quest became familiar with many of Liddell Hart's books. Since there was no regular army, and neither instructors nor pupils were experienced with units larger than a company, there was a natural tendency to concentrate too narrowly on practical training at the tactical level. Yadin found Rommel's book *Infantry Attack* ideally suited for instruction in fieldcraft, but he realized the need to stimulate thinking about war and strategy with a view to broadening the outlook of future commanders of a Jewish national army. For this purpose he found Liddell Hart ideal. It was not so much Liddell Hart the historian or the proponent of armour (Fuller also he regarded as too specialized a tank theorist), but rather Liddell Hart the general tactical theorist and, above all, the exponent of the strategy of indirect approach. In conversation Professor Yadin emphasized that books were all-important in these early years; he himself translated several chapters of Liddell Hart's *Strategy: the Indirect Approach* into Hebrew for the benefit of his students. He felt that Liddell Hart's value to the Israelis lay primarily on the strategic plane, citing for example the idea that the further from the front the enemy's line of communications could be cut, the greater would be the long-term effects. Furthermore, and this point was echoed by other Israeli commanders, Yadin had found Liddell Hart's books most useful in triggering off his own reflections about the organization and doctrine of the future Israeli forces. In his view there had later been a tendency to interpret Liddell Hart's ideas too literally and tactically. His own opportunity to implement a strategy of indirect approach would occur in the War of Independence, but by then he had disseminated ideas about the importance of this philosophy of war to dozens of young Haganah leaders who could not have read Liddell Hart in English.

Other interviews supplemented and in most cases corroborated this picture sketched by Yadin of Liddell Hart's influence on the Haganah before 1948. Former Prime Minister Rabin preferred to emphasize that Liddell Hart's doctrine of the indirect approach had largely coincided with Israel's choice of methods designed to overcome her inferiority in arms and numbers and the vulnerability of her people and territory. He thought that Liddell Hart had helped, however, in providing the Israelis with a theoretical justification and elaboration of their strategy. Yigal Allon, at that time Foreign Minister, confirmed that when he first met Liddell Hart in 1949 he could neither read nor speak English and had used the writer Jon Kimche as his interpreter. Allon had learned of his ideas before the War of Independence from translations and from talks with the veteran Haganah leader Yitzhak Sadeh, who on most issues was in agreement with Liddell Hart.

General Haim Laskov is another distinguished soldier who expressed a warm admiration for Liddell Hart, several of whose books he read before 1948. He was born in 1919 and served as a regular officer with the British Army in the Second World War. In the I.D.F. he displayed great versatility and originality. He was head of the Israeli Air Force between 1951 and 1953 and commanded the Armoured Corps in the Suez campaign. He was Chief of Staff from 1958 to 1960 and on leaving the Army became head of the Ports Authority. He is currently back in uniform as a sort of military ombudsman dealing with soldiers' complaints. General Laskov particularly admired Liddell Hart's ability to express tactical ideas in simple language suitable for instructing troops. He instanced the image of the 'man in the dark' and vigorously mimed the actions of the boxer who parries with one hand while punching with the other. He also thought that Liddell Hart had expressed better than any other theorist the importance of continuous movement in battle as distinct from an advance in separate phases. Employing exactly the same phrase as Professor Yadin, he remarked that he had found Liddell Hart's books most useful in 'triggering off' his own thoughts.

Only one foreigner is generally considered to have exerted a more direct and pervasive influence than Liddell Hart on Israeli military theory and practice—Captain Orde Wingate. Wingate served in Palestine from 1936 to 1939 on detachment from the British Army and his fervent Zionism and charismatic leadership made such an impact that he is still referred to as 'the friend' (Hayedid).[15] Wingate's outstanding practical achievement was to organize the Special Night Squads in 1938. These were small mixed units (nine in all) of British soldiers and Haganah men, which operated in northern Palestine in association with larger groups composed entirely of Haganah men. Under Wingate's leadership the S.N.S. became skilled at commando-style raids and ambushes, particularly in protecting the Iraq pipe-line to Haifa. His most important military legacy was to inspire the Haganah to adopt a more aggressive outlook in contrast to the defensive mentality which had naturally developed from the need to protect Jewish settlements. Politically he made no secret of his vision of the S.N.S. as the first step towards a Jewish National Army which one day he hoped to command.

Yigal Allon told the author that although he had not been a member of the S.N.S. he had commanded a unit of the Jewish Settlement Police which had taken part in three operations in Galilee under Wingate's direction. Much later, at an annual conference of the Institute for Strategic Studies at Oxford, Mr Allon had listed the three persons who had had the greatest influence on the I.D.F. as: first Wingate, second Liddell Hart and third Sadeh. In retrospect Wingate's contribution seemed all the more significant in that, as a non-Jew and a regular British Army officer, he had devoted himself—almost at the cost of his Army career—to the Zionist cause at a crucial phase in the struggle for national sovereignty. Major-General Avraham Yoffe, who was born in 1914, though later a keen reader and friend of Liddell Hart, also gave Wingate precedence as the major influence on his military outlook. He had served under Wingate in the S.N.S. and indeed was wounded in the largest battle at Dabburiya near Nazareth.[16] General Ariel 'Arik' Sharon is of a later generation than Yoffe and Allon: born in 1928, he has only boyish recollections of the Arab rebellion from 1936 onwards. Nevertheless his hero from those days was Wingate and he later read avidly about his exploits in Abyssinia and Burma. After serving in the ranks for several years, he attended an officers' course in 1950 where his chief instructor was Rabin. He had not then read anything by Liddell Hart and could only have heard him discussed by his seniors Allon, Yadin and Rabin. Mr Rabin's unequivocal opinion is that Wingate's influence was more important than Liddell Hart's in both practice and theory.

In later years Liddell Hart was pleased to recall that despite the apparently heavy odds against them, he had forecast an Israeli success just at the beginning of the War of Independence. He did so in an interview with David Kessler (Editor of the *Jewish Chronicle*), who felt anxious in view of the Arabs' superior numbers and modern weapons:

> I said that I did not think the danger would prove nearly as great in reality as it appeared on paper, and that the Jews should be able to repel the invasion, even though it came from several quarters—(1) because the Arabs were not good at combining; the different states were likely to play for their own hand, and thus dissipate their efforts; (2) because the Arabs were bad at administrative organization, and would be unable to maintain strong forces at such a distance from their base, over tenuous communications; (3) because the Arab was more careful of his own skin than the Jew—'the Arab didn't like being killed and the Jew didn't mind it'. That was the basic weakness that would handicap the Arabs' chances, and offset their numerical superiority.

Neither in this note, however, nor in an interview published soon after the War of Independence, did Liddell Hart suggest that Israel's best strategic solution might lie in a preventive strike; instead he spoke entirely in terms of a mobile defence to thwart the attackers.[17]

On 15 May 1948 the foundation of the state of Israel was proclaimed and the country was immediately invaded by the armies of Egypt, Iraq, Syria, Lebanon and the Arab Legion of the Kingdom of Transjordan. The incredibly confused operations lasted until the armistice imposed by the United Nations in January 1949. The war was waged sporadically in a series of separate combats and short campaigns interrupted on several occasions by truces enforced by the Security Council of the United Nations.

It is beyond the scope of this study to attempt to describe the operations.[18] The essential point is that at the outset the Haganah was little more than an irregular infantry force equipped only with light weapons and entirely lacking artillery, tanks and aircraft. When a tank 'battalion' was formed, only two of its six companies possessed tanks: one was equipped with ten French-built light tanks and the other with two Cromwells and a Sherman. In the early years of the I.D.F. armoured and mechanized units usually contained more half-tracks and a miscellany of vehicles with only a small proportion of tanks. By October 1948 the number of men and women in uniform exceeded eighty thousand (double the number in May) and centralized control had been more or less established by the General Staff. The Palmach, commanded by thirty-year-old Yigal Allon, had proved its capacity to fight sustained actions against regulars and was supplying commanders for improvised battalions and brigades—there were as yet no divisions. This date also marks a transition from hurried and scattered piece-meal actions to the concentration of large multi-brigade task forces to attack on one front at a time. First attempts were made to coordinate air strikes with ground attacks and even a small Navy came into operation. By the end of the year the guerrillas and home guards of the Haganah had become organized brigades of a national army under the direction of a Minister of Defence and a General Staff.[19]

Leading participants and more recent commentators have linked Liddell Hart's name with the Israelis' intellectual approach to the War of Independence. Yigael Yadin, the real Chief of Staff during the War, openly acknowledged that he had tried to direct operations in accordance with Liddell Hart's principles:

> There is no doubt that the strategy of indirect approach is the only sound strategy; but the constitution of the indirect approach in strategy—as brilliantly defined, explained and elaborated by Captain Liddell Hart—is far wider and more complex than in the tactical field. To exploit the principles of war for our purpose and base ourselves upon strategic indirect approach, so as to determine the issue of the fighting even before the fighting has begun, it is necessary to achieve the three following aims:
>
> (a) to cut the enemy's lines of communication, thus paralysing his physical build up;

(b) to seal him off from his lines of retreat, thus undermining the enemy's will and destroying his morale;
(c) to hit his centres of administration and disrupt his communications, thus severing the link between his brain and his limbs.[20]

Of course it would be erroneous to suggest that all Israeli commanders opted for the strategy and tactics of indirect approach. In fact there were several instances of frontal assaults on enemy strong points which had disastrous results. Five frontal attacks, for example, were made on the Jordanian-held fortress at Latrun; all failed and more than seven hundred lives were lost.

One brilliant operation does however merit special attention as an almost perfect illustration of Liddell Hart's theory on the strategic plane. This was Operation Horev (or Ayin) executed between 22 December 1948 and 7 January 1949 and designed to expel the Egyptians from the southern Negev. Its three chief executants were all directly or indirectly indebted to Liddell Hart's ideas: namely Yadin, Allon, the outstanding military leader of the war and by now in command of Southern Front, and his Chief of Staff, Rabin. The operation was also the one noteworthy achievement of the Israeli armoured unit, which otherwise performed rather poorly.[21]

The essence of the Israeli problem was that the Egyptians commanded all the important roads and were ready to meet an orthodox attack. They controlled the coastal road from El Arish to Gaza, the Auja-Hebron road as far as Bir Asluj and the two lateral roads Rafah-Auja and El Arish to Abu Agheila (see map). Yadin happened to notice that on an archaeological map the trace of a Roman road was shown running almost straight from Beersheba to Auja. He therefore sent a message to Rabin to reconnoitre the track to see if it was usable, and the latter cabled back 'difficult but passable'. With engineering improvements the route was made passable not only for half-track vehicles but even for medium tanks. As a result complete strategic surprise was achieved:

> While the Egyptian outposts at Bir-Asluj were eagerly watching the Beersheba road, expecting an attack from that direction, Israeli light forces emerged from the desert, captured a series of strongholds

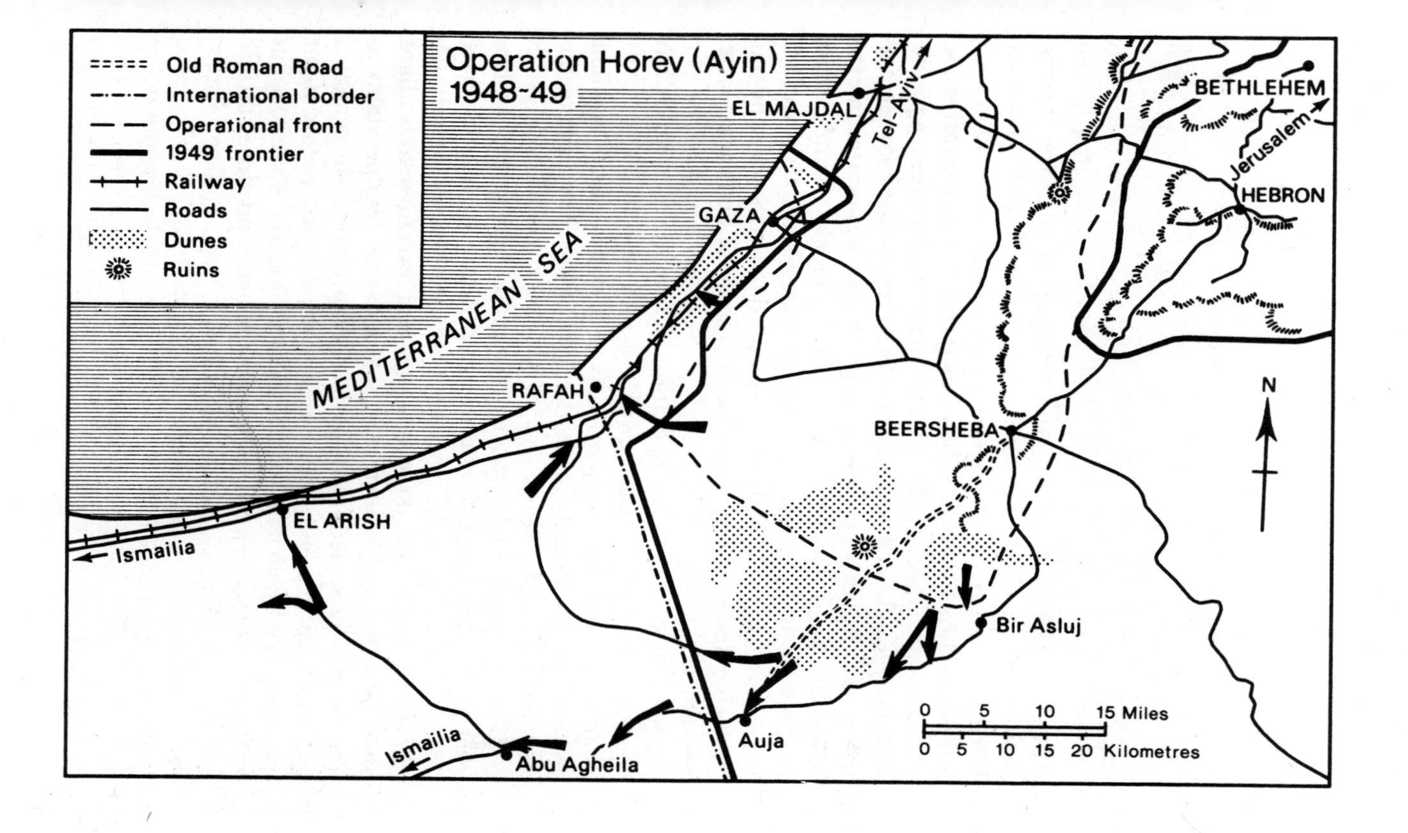
Operation Horev (Ayin)
1948-49
Old Roman Road
International border
Operational front
1949 frontier
Railway
Roads
Dunes
Ruins
MEDITERRANEAN SEA
EL MAJDAL
Tel-Aviv
BETHLEHEM
Jerusalem
HEBRON
GAZA
RAFAH
BEERSHEBA
N
EL ARISH
Ismailia
Bir Asluj
Auja
Abu Agheila
Ismailia
0 5 10 15 Miles
0 5 10 15 20 Kilometres

> further south, and blocked the Auja-Rafah road in two places. When Auja itself was attacked at dawn on the 25 December, it had already been cut-off both from its northern outposts and from its bases in the west. The garrison of that locality did its utmost, but after all reinforcements had been repulsed by the blocking forces on the road to Rafah it retreated into the desert in the early hours of the 27 December. A few hours later Bir-Asluj, now completely isolated, was occupied, and the Beersheba-Auja road was opened to our traffic, thus completing the first phase of our plan.[22]

After a brief interval to rest and bring up supplies the Israelis pressed on to take the Egyptian base at Abu Agheila on 28–29 December and an armoured column then swung north along the lateral road to El Arish. When this thrust was just short of the coastal road, however, Britain threatened to intervene and the Israelis were obliged to withdraw. Another column lay on the outskirts of Rafah when the Egyptians obtained an armistice on 7 January 1949. Despite these political obstacles the operation had been a brilliant success: the Egyptians had been expelled from the whole of Palestine except the Gaza strip by an attacking force barely superior in numbers and inferior in equipment.

Questioned about this operation by the author, Mr Allon replied in the spirit of Clausewitz's dictum cited in the previous chapter: he had had no particular textbook in mind but had worked out his movements in the light of his knowledge of the enemy forces and terrain.

Like the proverbial letter-writer who is told to his surprise that he has produced 'literature', said Allon (perhaps echoing Molière's M. Jourdain), so he had acted more or less instinctively and then discovered he had executed a brilliant indirect approach! Liddell Hart told him that his operations in the Negev provided near perfect examples of the indirect approach on a small and a large scale. He added that night fighting even by such large formations as brigades was also in his opinion a component of indirect approach, and this skill was traceable to Wingate rather than Liddell Hart. Professor Yadin by contrast was quite ready to acknowledge that he was influenced by Liddell Hart's general thinking in his conduct of operations in 1948–49. Without wishing to detract from Allon's outstanding achievements as a field commander, he thought that Rabin's brilliant analytical mind was chiefly responsible for the planning of 'Horev' and other operations on the Southern Front. Rabin had not read much then and only knew of Liddell Hart indirectly through his (Yadin's) instruction.

Within a few months of the ending of the war in 1949 the Army of Independence had virtually disappeared. The officer corps suffered even more severely than the other ranks, partly because few had regarded themselves as career professionals, but also because of political pressure. Prime Minister Ben Gurion (of the Mapai Party) conducted a relentless political campaign against the Palmach, many of whose officers belonged to the left-wing Mapam Party. Israeli soldiers, unlike the British, made no pretence of being above party politics; indeed Palmach training inculcated a collectivist ideology of left-wing Labour Zionism. Ben Gurion therefore had some grounds for regarding Palmach military heroes, such as Yigal Allon, as political rivals. In the event Palmach was disbanded and the Army lost two out of four front commanders (including Allon), six out of twelve brigade commanders and many more experienced officers.[23]

Yigael Yadin, who succeeded the ailing Jaacov Dori as Chief of Staff in November 1949, was a non-political officer who commanded the respect not only of Mr Ben Gurion, but also of the Palmach, Haganah and British-trained officers. Yadin faced an extremely difficult task. He was given the political directive of combining severe economies with instant preparedness for war. His organizational problem was to form regular brigades when he had hardly any experienced senior commanders to call upon. In social terms he had to offer sufficient inducements to attract a cadre of first-class regular officers without having the money to compete with business and commerce. For all these reasons he attempted to introduce some of the features of a regular army like the British, such as establishing officers' clubs and differentiating more clearly between officers and other ranks. He also strove to achieve a blend between the few surviving ex-Palmach officers such as Rabin, and former British officers such as Ben Artzi (head of the Quartermaster Branch 1950–52) and Laskov (head of the Instruction Branch 1948–51). When the financial stringencies proved too great a problem and Yadin retired to take up an academic career, there was a reversion to the more free-and-easy egalitarian atmosphere of the Palmach but without its political overtones. His lasting achievement was, however, to create the reserve system which has endured essentially unchanged to the present.

As head of the Instruction Branch during the crucial period when Yadin was establishing the foundations of the I.D.F., and later in 1956 as Commander of the Armoured Corps, Haim Laskov was to play a vital role in introducing Liddell Hart's ideas. He shared with Yadin and Allon a firm belief in the value of the indirect approach as a guide to strategic thought, but unlike them he was also a keen student of Liddell Hart as the proponent of tanks. Thus despite the poor performance of Israeli tanks in 1948 and the fact that in 1950 there were still only a few dozen—mostly Sherman M4s—Laskov and his planning group assigned armour the key role in ground combat. In suggesting that tanks could best be used in Liddell Hart style, that is, independently with mechanized infantry for deep penetration into the enemy's rear areas, they were too advanced for what was still basically an infantry army. As Chief of Staff from 1953 to 1958 Moshe Dayan was at first distrustful of tanks, preferring fast-moving infantry battalions carried by jeeps, half-trucks and armoured cars. At the beginning of the Sinai operations in 1956 he accordingly relegated armour to an infantry support role. (In fact the tanks were to follow on transporters and their crews in buses.) This was because he equated battlefield mobility directly with vehicle speed. Tal and his followers later redefined it as the ability to move in the face of hostile fire, with the result that the battle tank became the dominant weapon in their eyes.

Though he was never a disciple of Liddell Hart's, General Dayan's strategy in the 1956 Sinai campaign was very much in accord with the spirit of the former's doctrine. Rather than a direct attack on the Egyptians' fortified perimeters with the object of destroying their army, Dayan believed that the enemy's forces would disintegrate if the Israelis could penetrate deep into Sinai and cut their communications. His stated intention was 'to confound the organisation of the Egyptian forces in Sinai and bring about their collapse'.

The Israelis began the war by dropping a small paratroop force deep behind the Egyptian front lines near the eastern entrance to the Mitla Pass, thus suggesting that an attack on the Canal was intended and at the same time providing a pretext for

Anglo-French intervention. Moreover, instead of first attacking the nearest objective to the frontier, the Gaza strip, the Israelis left this till last, first securing the crossroads at Abu Agheila and Umm Katef controlling the routes westward to the Canal. Meanwhile further south Sharon's 202nd Brigade was to follow 'the line of least expectation' by driving south-west into the desert to reinforce the paratroops holding the entrance to the Mitla Pass. Dayan took grave risks in sacrificing proper preparations in order to achieve surprise; some reservists went into battle only a few hours after reaching the jump-off position.

It was, however, primarily due to the commander of the Armoured Corps, Haim Laskov, that tanks were to play a vital part in the operations very much along the lines advocated by Liddell Hart. In pre-war planning Dayan at first assigned Ben Ari's 7th Brigade (the only armoured brigade in the standing army) a purely diversionary role on the Jordanian front and, when Laskov protested, relented only to the extent of allotting the brigade to a support role for the infantry in the main attack on the Abu Agheila-Umm Katef crossroads. Laskov, as an avowed follower of Liddell Hart, wanted the tanks to be concentrated in a few all-armoured units which would be sufficiently powerful to break through the enemy's defended zone, thrust deep into his rear areas, and then fan out to threaten several objectives.

In the event the Southern Front Commander, Assaf Simchoni, decided to flout Dayan's order and allowed 7th Brigade to go into action twenty-four hours ahead of schedule. This entirely upset G.H.Q.'s deception plan, by which Sharon's brigade was to have a two-day start on the main offensive, but it nevertheless resulted in a spectacular victory. While the infantry was held up on the Umm Katef section of the perimeter, Ben Ari's tanks infiltrated into the rear of the Egyptians at Abu Agheila and then, with hardly a pause, drove on westward in *blitzkrieg* style reminiscent of Guderian's dash through France in May 1940, to take control of most of Sinai by the end of the second day (31 October). Thus, as Luttwak and Horowitz sum up, 'the outstanding success of 1956 was the result of a classic armoured attack—in direct contradiction to Dayan's orders, and all the plans of G.H.Q.'[24]

An equally flagrant disobedience of G.H.Q.'s orders occurred

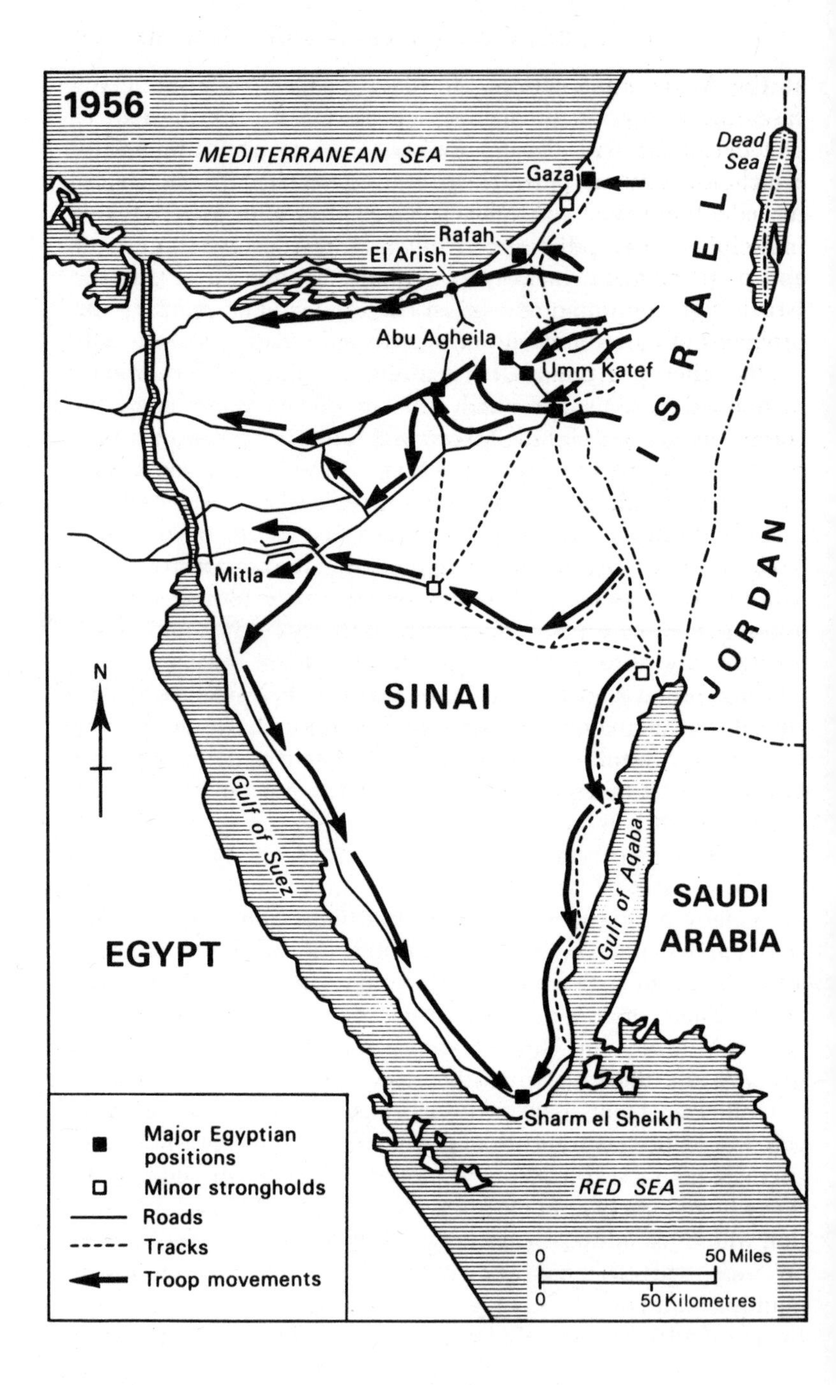

1956
MEDITERRANEAN SEA
Gaza
Rafah
El Arish
Abu Agheila
Umm Katef
Dead Sea
ISRAEL
JORDAN
Mitla
SINAI
N
Gulf of Suez
Gulf of Aqaba
SAUDI ARABIA
EGYPT
Sharm el Sheikh
RED SEA
Major Egyptian positions
Minor strongholds
Roads
Tracks
Troop movements
0
50 Miles
0
50 Kilometres

at the Mitla Pass, where Sharon's brigade suffered heavy casualties as a consequence of pushing on into the pass on 31 October, again with the connivance of the Front Commander, Simchoni. Dayan, perhaps literally as well as figuratively, turned a blind eye to what he characteristically termed 'positive indiscipline', i.e. indiscipline stemming from an excess of drive and pugnacity; but in the *post mortem* it became clear that the Israeli high command had taken risks by virtually ignoring the problems of co-ordination and control.[25]

One other spectacular exploit in the 1956 campaign should be mentioned since its commander, Colonel Avraham Yoffe, later corresponded with Liddell Hart. This was the cross-country march of 9th Infantry Brigade to Sharm el Sheikh, which was all the more creditable in that Yoffe and his men were reservists mobilized only on the second day of the war. In an interview Yoffe, a bulky, jovial character whose command of English is still imperfect, remarked that no one expected a whole brigade accompanied by a convoy of more than 200 vehicles to attempt a route only fit for camels. There was only one oasis en route. The Egyptians had assumed that his reconnaissance company constituted the whole force and had therefore been completely surprised. Sharon's brigade was simultaneously advancing down the east bank of the Canal and the two forces took the objective with minimal opposition. Yoffe's brigade was said by a journalist to have 'made the impossible across the impassable'. Not all the reserve brigades did so well.[26]

In less than eight days' fighting the Israelis had routed the equivalent of two Egyptian divisions and conquered an area about three times the size of their own territory and all at a cost of less than 200 troops killed. Nevertheless as a result of the ceasefire imposed by the United Nations their security was scarcely improved at all. Consequently their strategic doctrine remained unchanged in essentials. It was assumed that the next war would be short and that Israel could not afford to let her enemies take the initiative. Victory would probably go to the side which launched an all-out pre-emptive strike. This assumption was given added significance by the massive arms build-up on both sides in the years after 1956, Egypt being supplied mainly by the Soviet Union; and Israel by France, Germany and Britain. Indeed it was becoming clear by the end of the 1950s that the arms race would transform the nature of the Middle East conflict. This was most obviously true in the air, since both sides were buying the latest types of fighters and bombers, such as the French supersonic Super-Mystère fighters acquired by Israel in 1959–60 and the Soviet MIG-21s and Tu-16 bombers supplied to Egypt. Time would show that the Israelis' technical and educational superiority still gave them the edge in absorbing ever more sophisticated weapons. Equally important, the Israelis shopped carefully, buying weapons selectively and insisting on modifications to suit their own situation and doctrine, whereas the Egyptians appear to have accepted whatever the Soviet Union chose to supply and adopted Soviet tactics together with the equipment.

Not only was the technological situation transformed in the decade following the Sinai campaign, but the Israelis also persevered with their system of the very rapid turnover of senior officers, no matter how successful. Dayan deserves most of the credit for this unorthodox system whereby able officers obtain positions of responsibility much younger than in other countries and then retire to take up another career. This does not necessarily mean that younger men have better ideas but, in addition to physical and mental fitness in the commander, it does ensure that strategic and tactical doctrines are constantly under critical scrutiny.[27]

By 1956 the two most distinguished proponents of the Liddell Hart doctrine of indirect approach from the war of Independence, Yadin and Allon, had left the Army, the former for an academic career and the latter for politics. The two commanders who became most closely associated with him from the later war were Generals Sharon and Laskov.

Liddell Hart first met Sharon at his home, then at Wolverton in Buckinghamshire, in August 1958 when the latter was taking the course at the Staff College, Camberley. In the following autumn Sharon wrote several times asking for Liddell Hart's comments and further suggestions for reading on the topic he had chosen for his main assignment; namely decision-making in the British and German armies during the Second World War. Sharon wanted to know in particular whether Liddell Hart approved of Rommel's highly individualistic style of leading from the front which might well have served as a model for his own method. Liddell Hart at first attempted to satisfy him with a brief 'terribly busy' reply, but Sharon persisted until he was given a detailed account of commanders' positions during the North African campaign. In his letter of thanks on 15 October 1958 Sharon wrote:

> I was brought up in the Israeli Army which, no doubt, was very much influenced by your unorthodoxic [*sic*] school of thought, and of course I am strongly in favour of your ideas.

Sharon visited States House, Medmenham in January 1963 and was impressed by his host's great reservoir of military knowledge on almost every topic. They met for the last time in July 1968, when Sharon presented Liddell Hart with a signed photograph of himself for the 'rogues gallery'. Despite the mutual admiration that existed between the two, it would be straining the evidence to list Sharon among Liddell Hart's 'disciples' except by the most tenuous criteria. The former had proved himself an aggressive, ruthless and daredevil leader with an instinctive flair for tactics from the early 1950s when he commanded the élite 101st paratroop battalion. He is not, by comparison with Allon or Laskov, let alone Yadin, a scholarly soldier by nature but essentially a brilliant, charismatic field commander who, in the 1967 campaign particularly, was to display operational talents of a very high order. Doubtless his correspondence with Liddell Hart when at Camberley served to confirm his own predilection for front-line leadership, but on meeting him one quickly realizes that he is an extremely self-confident soldier who goes his own way regardless of the consequences. It is impossible to think of him as anyone's 'disciple'.[28]

Haim Laskov on the other hand, though by no means lacking in strength of personality, is very willing to testify to Liddell Hart's influence on his own thinking. He is probably the only important Israeli soldier to have drawn more on Liddell Hart's ideas about armoured warfare than on his general theory of the indirect approach.[29] They first corresponded in December 1956 when Liddell Hart wrote to congratulate him on his brilliant performance in command of the Armoured Corps in the recent war. In 1959, when Laskov was Chief of Staff, Liddell Hart sent him a copy of *The Tanks* with a flattering inscription. The recipient replied that the volumes were 'an inspiration to any student of war and weapons'. Their subsequent correspondence is largely concerned with tank capabilities and problems of mechanized warfare.

It was also Liddell Hart's good fortune that such a thoughtful and like-minded general was Chief of Staff at the time of his sole visit to Israel in 1960. As a former head of the Air Force and ex-commander of the Armoured Corps, Laskov was probably the ideal successor to Dayan as Chief of Staff in that he sought to reshape the I.D.F. for the 1960s on the basis of fast-moving armoured forces supported by a powerful Air Force. According to Luttwak and Horowitz his publications (chiefly in the specialist Hebrew military journals) 'form the most comprehensive statement of Israeli military doctrine' (for the 1960s). 'His ideas', they continue, 'went far beyond a commonplace striving for tactical mobility'; he and his successors 'grasped the deeper implications of mobility, and set out to shape the Army's organization, doctrines and method of command around the sometimes subtle requirements of mobile warfare'.[30] Although the evidence is indirect it seems reasonable to associate Liddell Hart's name with Laskov's doctrine; the former would certainly receive unambiguous tributes after the theory had been brilliantly implemented in 1967.

Liddell Hart's visit to Israel from 20 March to 5 April 1960 as a guest of the Government was something of a triumphal tour. He was hailed by the Press as 'the greatest military expert of our time', 'the Clausewitz of the 20th Century' and 'the Pope of theory', while he in return was quick to laud his hosts as his best disciples and the most brilliant soldiers in the world. He had interviews with all the 'top brass', including discussions over the map with Laskov and his Vice-Chief, Rabin, and with the Prime Minister, Ben Gurion. He also gave five public lectures which some people found difficult to follow because of his indistinct delivery.[31] Journalists were clearly puzzled by his appearance, which in no way coincided with their image of the bluff, red-faced, thickset British officer. According to one interviewer:

> Liddell Hart was, in his appearance and manner of speaking, far

> different from what we had imagined as the greatest military expert of our time. He seemed more like an English gentleman whose sole warlike occupation had been a single lion hunt in his youth, and who since then was accustomed to recount this experience in his exclusive club.

Another journalist delightfully characterized his somewhat incoherent manner of speaking:

> He is also a very charming man. In spite of his 65 years, he still stands erect, showing his full height, and what a height—almost 2 metres . . . A gold watch chain decorates his waistcoat, a pipe never moves from his mouth and he speaks—but that's a story in itself—he swallows every word, skips half a sentence and finally—repeats himself. That is the famous Cambridge style of speaking, a vestige of the Golden Age of the British Empire. If you are not English, then it's really not important whether you understand me or not. The story is told that somebody once suggested to a Defence Minister in one of the western states that he invite Liddell Hart as a military adviser. 'In the atomic age, that's impossible,' the Minister answered, 'by the time I've understood what he's advising me to do, we'll already have lost the war.'[32]

One suspects that on this brief but hectic tour the Israelis gained less from Liddell Hart's public pronouncements than from his genius at stimulating individuals and provoking them to fresh thought. This was certainly the impression of one of his most brilliant disciples, Dr Israel Beer, a leading military historian and commentator who, in 1961, was found guilty of espionage on behalf of the Soviet Union. Beer, who had chaired one of Liddell Hart's lectures, wrote a few days after his departure:

> I think that visit and your lectures have done a lot of good to our officers and the results will be felt in our Army sooner or later. You have forced many people to use their brains again, people who have become lazy, too spoiled and arrogant by successes, whose real causes they do not completely understand. You have by your work during your whole lifetime tried to infuse 'old armies' with dynamism and a flexible way of thinking. Believe me, that a 'young army' like ours, does even need more what you have to give, such as sound reasoning, penetrating analysis and above all your moral stature.[33]

The Six Day War in 1967 probably provided the most brilliant examples of the indirect approach in Liddell Hart's lifetime. For the historian of the Israeli forces all the fronts are of equal significance, but for the student of Liddell Hart's ideas Sinai is once again the crucial theatre of operations.

On that front the aim of the newly appointed Minister of Defence, Moshe Dayan, was to destroy the Egyptian army—an objective which in his earlier career Liddell Hart would have criticized as Clausewitzian.[34] The Israelis gained a decisive advantage by a pre-emptive air strike, destroying more than two hundred aircraft on the first day, as against a total loss of less than fifty of their own during the war. On the ground three divisional task forces were ordered to prise open the formidable Egyptian perimeter defences before driving across Sinai. In the north Tal's division was to break through the Rafah defences and advance to El Arish. In the centre Yoffe's smaller division, comprising only two reserve armoured brigades, was detailed to advance through difficult desert terrain and cut the roads leading to El Arish, thus shielding Tal's flank. Perhaps the most difficult task, that of enveloping the deep Egyptian defences round the crossroads at Abu Agheila and Umm Katef, fell to Sharon's division.

Tal's attack did not go according to plan but succeeded beyond all reasonable expectation through improvisation by his unit commanders and sheer hard fighting. When the tanks of his leading battalion secured a breakthrough they drove on for several miles amidst bewildered Egyptian defenders without waiting for infantry support. El Arish was captured in just over half the time allotted. While Tal's forces were advancing along the Rafah-El Arish road and Sharon's were preparing for a night attack on the central crossroads, one of Yoffe's brigades struggled through the sand dunes between them. This was a model illustration of Liddell Hart's concept of indirect approach since, in taking 'the line of least expectation', it sought to achieve surprise by overcoming natural rather than human resistance. In other words, Yoffe preferred 'a day-long struggle through soft sand' rather than risk a frontal attack. The operation bore a close resemblance to Yadin's use of the forgotten Roman road in the same theatre in 1948. Yoffe told the author that the most important influence on his decision was

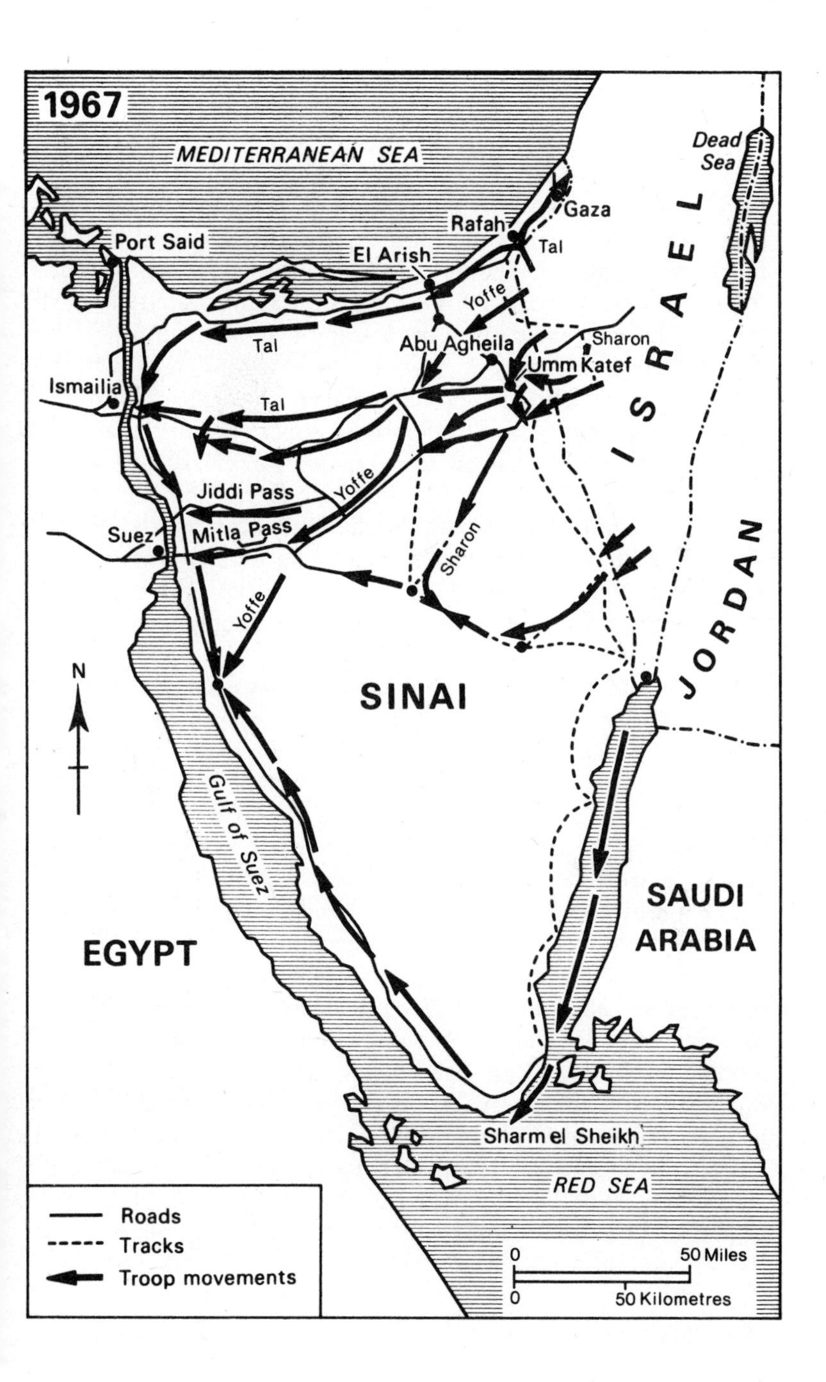

1967
MEDITERRANEAN SEA
Dead Sea
Gaza
Rafah
Tal
Port Said
El Arish
Yoffe
ISRAEL
Tal
Abu Agheila
Sharon
Umm Katef
Ismailia
Tal
Yoffe
Jiddi Pass
Suez
Mitla Pass
Sharon
JORDAN
Yoffe
N
SINAI
Gulf of Suez
SAUDI ARABIA
EGYPT
Sharm el Sheikh
RED SEA
Roads
Tracks
Troop movements
0
50 Miles
0
50 Kilometres

his experience of similar treks in his time with the 8th Army in the Second World War. He knew the terrain well and cleared the plan with the Southern front commander, General Gavish, and the Chief of Staff, Rabin. In a march of some twenty miles his brigade had encountered only one Egyptian company. His route took him behind the main enemy defences and enabled him to cut off their reinforcements to the front, where Tal's infantry units were heavily engaged, having failed to push through the gap opened by his leading armoured units.

Yoffe's outstanding feat, however, was still to come. On the third day the brigade which had trekked across the sand dunes (Colonel Shadmi's) was ordered to advance more than one hundred miles through enemy territory to block the eastern entrance to the Mitla Pass. This it achieved with only nine tanks left, because of lack of fuel, and with a little infantry and mortar support held out for eighteen hours against increasingly desperate Egyptian attacks. The entrance to the pass became a killing ground where Egyptian tanks and infantry were annihilated.[35]

Sharon's victory at the Abu Agheila-Umm Katef crossroads was probably the most brilliant of all because he chose to operate by night against well-organized Egyptian defences in depth consisting of minefields and entrenchments supported by tanks and artillery superior to that of the attackers. Whereas Tal's success in the northern sector was achieved by improvisation, Sharon's was a model of precise co-ordinated planning in which paratroopers led the way to clear a path through the minefields supported by a tremendous pin-point barrage by self-propelled artillery and mortars, and followed up by two tank forces which, after infiltrating the defences, converged from east and west. The least error of timing would have jeopardized the whole operation and might well have resulted in Israelis firing on each other. Sharon by this operation proved himself a master of set-piece tactical planning as well as a dynamic leader in battle. The lessons he drew from this experience were that every staff officer must be committed to the success of the plan and that the commander must stay forward and intervene in the battle as he thinks fit. In his view, as he demonstrated again in 1973, coordinated actions on a large scale cannot be controlled from a rearward headquarters.[36]

By these actions the Israelis not only disrupted the Egyptians' frontier defences but also ruined their carefully prepared counter-offensive plan by which the attackers were to be rendered immobile by the 'shield' of mixed infantry, artillery and tanks and then destroyed by the 'sword' comprising the equivalent of two armoured divisions. But the Israelis' supreme achievement lay in the manner in which they exploited their initial success by driving through the Egyptians' rear areas to block all exits through the Mitla and Jiddi Passes and across the canal. In a much-publicized article in *Encounter* Liddell Hart, with pardonable hyperbole, described this second phase as the subtlest and most effective application of the 'indirect approach' in the record of modern warfare. The Israeli strategy, easy to describe but extremely difficult to implement, was to drive *through* the retreating Egyptians—in some cases along the same road—and then set up blocks which the disorganized enemy would have to attack and overwhelm in order to escape. By the night of 8 June, the day before the United Nations ceasefire was announced, the Israelis had achieved the utter rout of the Egyptian army in Sinai, a much larger force than Rommel had commanded at the battle of El Alamein.[37]

Several Israeli soldiers and academics speedily paid tribute to Liddell Hart as the chief intellectual inspiration behind the strategy and tactics of the Six Day War. Dr (now Professor) J. L. Wallach of the University of Tel Aviv (also a Colonel in the reserve) wrote that the Israelis had been worthy disciples of the 'indirect approach' for the third time:

> And that, above all, not in the technical concept alone, but mainly in the mental and intellectual sphere, by the intelligent selection of the time, direction, method and power of our moves, especially on the strategic level.[38]

Haim Laskov commented that:

> To my knowledge, your ideas on Armour in their most versatile role were put to the test by Tal's Division in the sector of Rafah-El Arish . . . Not only was the terrain difficult but resistance was toughest.
>
> I do have a feeling that on Armour you sit back, chuckle and wonder what fate and the Israelis did to vindicate a life-long teaching and preaching.

In his reply to General Laskov, Liddell Hart remarked that he had always had mixed feelings regarding German generals' tributes to his influence on their army: 'But there have never been any such mixed feelings, and only pleasure, in seeing how the leaders of Israel have applied those ideas, even still better.'[39]

In his *Encounter* article Liddell Hart emphasized that, generous though Guderian and Rommel were in their acknowledgements of his influence, they had not grasped the subtler side of the indirect approach so well as the Israelis. The latter indeed provided a rare exception to the experience of history that armies learn only from defeat, not victory. In the recent campaign they had improved on their performance in 1956 by the skilful combination, in Sinai, of the strategical offensive with the tactical defensive. 'You and your countrymen,' he wrote to Laskov, 'have certainly been my best disciples.'

Perhaps the most gratifying tribute came from Professor Yigael Yadin in commenting on Liddell Hart's review of the Churchills' book on the Six Day War:

> I can see from your remarks how the authors completely missed the basic strategic principle of Rabin's plan (and it was *his* plan, which was only slightly modified by his political superiors), i.e. the subtle and superb exploitation of the true strategy of indirect approach. All of us, as you know—Rabin not least—have learnt and re-learnt for many years your books, and the success of the six day battles was in a way a tribute to the soundness of your teachings which, thank God for us, were not learnt by the Arabs, including the British-conceived Arab Legion. The latter obligingly moved into the trap although the whole campaign against them was a brilliant chain of tactical improvisations all based on the main doctrine of the indirect approach.[40]

Only a nation with an unbroken record of victories and such high military standards as the Israelis would have regarded the outcome of the 1973 war as in any sense a *military* defeat. Despite the acrimonious *post mortem* which they have conducted since, it is by no means obvious to a detached outsider that they still enjoyed the option of a pre-emptive strike.[41] The immense strategic gains from the Six Day War had strengthened the Israelis' military position but undermined their diplomatic leverage and moral advantage. In short, they were no longer generally regarded as underdogs and were now viewed by some former sympathizers as aggressors. Moreover, while in many respects they had continued to learn lessons even from the crushing victories of 1967, a certain amount of complacency and even hubris was apparent.[42]

Liddell Hart was of course no longer alive to comment on the outcome of the War of Atonement, but it seems reasonable to assume that he would have criticized the static defensive element symbolized by the construction of the Bar Lev Line, which was adopted against the advice of the more offensive- and mobile-minded General Sharon. Also, since he had stressed the necessity ever since the 1920s for mechanized infantry to work in close cooperation with armour, it is safe to assume that Liddell Hart would have criticized the post-1967 reorganization associated with the name of General Tal, by which the tank battalions in each brigade were largely deprived of their mechanized infantry and mortars. While it would be over-simplifying to talk of 'pure, all-tank formations', the opening battles of 1973 in Sinai clearly demonstrated that the Israeli armoured units were very inadequately supported against the Egyptian defences, now devastatingly equipped with the latest anti-tank weapons.

In fact it was the Egyptians who more nearly fulfilled the spirit of Liddell Hart's teaching in the opening phase by combining their strategic offensive across the Canal with a strictly tactical defensive. Infantry formed the basis of their defensive shield, armed with the wire-guided anti-tank Sagger missile, effective at ranges which prevented the Israeli tanks from coming into action. The bulk of the Egyptian armour was held back on the west side of the canal pending the destruction of the Israeli armoured units. So far so good, but the Egyptians showed themselves to be sadly deficient in Liddell Hart's other precepts of offensive mobility and flexibility. Having won the opening round, the Egyptians proved incapable of exploiting their success by advancing and so allowed time for the Israelis to recover and regain the initiative.

In the final phase of the war, interrupted by the ceasefire imposed by the U.N. Security Council, the Israelis—and particularly Sharon—showed that they were still masters of the mobile counter-attack. Sharon's strategy after crossing to the west bank of the Canal was in the boldest tradition of Guderian and Rommel. With an advance guard of only some thirty tanks and two thousand infantry he scorned the obvious solution (and his orders) to remain on the defensive until the bridgeheads on both sides of the crossing were secured, and by fanning out in attack created confusion throughout the rear areas of the Egyptian 2nd Corps.[43] There has indeed seldom been a more perfect example of Liddell Hart's ideas of deep penetration and the paralysis of the enemy command by cutting communications, threatening alternative objectives and spreading panic. But for the Russian intervention it can hardly be doubted that the Egyptians, after a promising start, would have suffered a defeat at least as calamitous as in the Six Day War. Much of the combat in this war, as for example on the Golan Heights, can hardly be interpreted, even tenuously, in terms of the indirect approach. With the growth of the defensive power—and self-confidence—of the Arab nations it seems unlikely that the Israelis will have another opportunity to implement the strategic indirect approach on the ground comparable to their successes in 1967 and 1973.

As for an eventual political settlement, Liddell Hart's generally pessimistic and very cautious reflections after the 1967 war have not to date been disproved in essentials. He felt that the history of the last half-century in the Middle East showed that little could be expected from negotiations; the only hope lay in time and stability. Writing before the oil crisis, he remarked that it would be foolish of the Western Powers to try to hasten a settlement either from altruism or from selfish concern for their own short-term interests. Experience of Arab politicians offered no ground of hope for conciliation. Israel must retain control of all, or almost all of the ground she had conquered. He went further by suggesting in an interview that the Israelis would have been well advised to have pressed further into Syria in order to establish an independent Druse state which could serve as a buffer between Jordan and Syria.[44] Thus, although he by no means ignored the political complexities of the Middle East situation, he was first and foremost a strategic commentator with no particular insights into the all-important issue of how military victories could be translated into a permanent settlement.

* * *

In trying to answer the extremely difficult question of how much influence Liddell Hart had exerted on Israeli military theory and practice, this chapter has indicated that while it would be foolish to take the simplistic claims of certain journalists literally, it would be even more unjust to dismiss the formidable evidence in his support. In short, Liddell Hart received as much credit and praise for Israeli military achievements in 1948–49, 1956 and 1967 as any theorist, and particularly a foreigner, has the right to expect during his own lifetime. Doubtless had he still been alive in 1973 further tributes would have been forthcoming and possibly another signed photograph or two to join those of Yadin, Allon and Sharon in his study. The main point seems to be that his influence was mostly implanted indirectly, that is by means of his books and articles, and most officers would meet these in translation or at second hand from lectures. In the 1940s Yigael Yadin was the crucial medium who transmitted Liddell Hart's broad thinking about war to a small but highly significant group of officers, several of whom, such as Rabin, reached key positions. In the 1960s a later generation of officers had the importance of the strategy of indirect approach impressed on them in university courses, in particular those of a wholehearted follower of Liddell Hart at Tel Aviv, Professor J. L. Wallach. As regards armoured warfare, Liddell Hart's ideas were expounded chiefly by General Haim Laskov in the mid-1950s and early 1960s.

Ultimately, it can be argued, all physical achievements such as military victories are determined by thought, but there is usually a long and complex process of germination between implantation of the written or spoken word and the successful executive decision. It is as well to remember in attempting to compare the influence of the pen and the sword that Liddell Hart himself recognized that the Israeli commanders had not merely fulfilled his principles but improved upon them. Finally, unlike the writer, the general is likely to be dismissed and sometimes even dishonoured if he fails; it is therefore only fitting that he should receive the lion's share of the credit if he is successful.

Notes

1. His visit to Israel was said to have received more press reportage than any since that of Marilyn Monroe. According to the list kindly supplied to me by General Laskov, none of Liddell Hart's articles published in Hebrew in the military journal *Ma'arachot* concerned tanks or technical matters; all were about policy and strategy.

2. Edward Luttwak and Dan Horowitz *The Israeli Army* (Harper: Allen Lane 1975) pp. 205–206 (henceforth referred to by its title). I am heavily indebted to this excellent study throughout this chapter. The authors suggest that Sharon's advance on the west bank of the Canal in 1973 was delayed by the overriding need to minimize casualties.

3. Liddell Hart 'Israel's Strategic Position' in *The Jewish Standard* 29 April 1949, and 'Strategy of a War' in *Encounter* February 1968.

4. See, for example, Yigal Allon *The Making of Israel's Army* (Vallentine Mitchell: Universe 1970).

5. Lest this paragraph appear too eulogistic it should be mentioned that the Zionist Jews have impressed even some sympathizers as excessively legalistic, prone to complain and incorrigibly factious—see Christopher Sykes *Orde Wingate* (Collins: World 1959) pp. 154–155.

6. The Palmach was a full-time mobile striking force of the Haganah established in 1941.

7. Laskov, for example, was recalled from his studies at Oxford to become Deputy Chief of Staff in 1955. As a reservist Yoffe commanded a brigade in 1956 and a division in 1967.

8. *The Israeli Army* op. cit. p. 79. Compulsory service applied to women from the start, though for a shorter period than men and with various exemptions.

9. Ibid. p. 203.

10. Interview with Mr Allon 19 September 1975. Liddell Hart in *Encounter* op. cit. Though the Arabs living within the pre-1967 borders of Israel (numbering about 400,000) are certainly Israeli citizens, they have not generally been conscripted for military service.

11. Liddell Hart *Memoirs* II pp. 181–183. Liddell Hart Papers 13/80 typed interview transcripts pp. 5 and 19. For a reference to Liddell Hart's meeting with Ben Gurion in 1936 see the former's published interview with David Kessler in *The Jewish Chronicle* 22 July 1960.

12. According to C. Sykes, on the other hand, Wingate did not like to be reminded of his kinship with or resemblance to T. E. Lawrence: see *Orde Wingate* op. cit. p. 133.

13. *The Israeli Army* pp. 17–22.

14. Interview with Professor Yadin in London 27 October 1975. Interview with Professor J. L. Wallach 17 September 1975.

15. Wingate's nickname of 'the friend' was at first employed sarcastically: see *Orde Wingate* p. 114.

16. Ibid. p. 157.

17. 11/1948/11 Diary note 22 May 1948; and 13/80 for interviews published in *The Jewish Standard* 29 April and 13 May 1949.

18. For a thorough account of the military aspects see Netanel Lorch *The Edge of the Sword* (Putnam 1961).

19. *The Israeli Army* pp. 36, 45–48, 54–55, 63–64.

20. Yigael Yadin's contribution to Liddell Hart's *Strategy: the Indirect Approach* (1954 ed.) p. 387.

21. *The Israeli Army* pp. 68–69.

22. Interview with Professor Yadin; and his contribution to *Strategy: the Indirect Approach* pp. 396–397.

23. Interview with Yadin. *The Israeli Army* pp. 70–74.

24. Ibid. pp. 143–153. Moshe Dayan *Diary of the Sinai Campaign* (Weidenfeld: Harper 1966) p. 210. Dayan became a convert to tanks as a result of the 1956 campaign.

25. *The Israeli Army* pp. 158–160. But General Sharon, when interviewed in September 1975, told the author that after Laskov had investigated his breach of orders it was Mr Ben Gurion rather than Dayan who exonerated him.

26. *The Israeli Army* pp. 156–157. Interview with Avraham Yoffe 18 September 1975. For a graphic and humorous account of his march to Sharm el Sheikh see Robert Henriques *100 Hours to Suez* (Collins: Viking Press 1957) pp. 114–135.

27. Since the 1973 war the I.D.F. has possibly accumulated an embarrassing number of senior officers 'in the wings' but available for active employment. They include Generals Sharon, Tal, Gonen and Laskov.

28. *The Israeli Army* pp. 116ff. Interview with General Sharon 18 September 1975. Sharon to Liddell Hart 16 September, 29 September, 15 October and 13 November 1958. Liddell Hart to Sharon 10 and 13 October 1958.

29. In a letter to Laskov dated 2 May 1969 Liddell Hart wrote that he had seen it reported that Ben Gurion had declared that it was Laskov in particular who took up and applied his theory of the use of armoured forces. Laskov's reply (on 11 May 1969) appeared to confirm Ben Gurion's statement.

30. *The Israeli Army* p. 172. Interview with General Laskov 17 September 1975.

31. Interview with Yitzhak Rabin 18 September 1975.

32. 13/80 Typed interview transcripts pp. 8, 21.

33. Ibid. Israel Beer to Liddell Hart 15 April 1960.

34. In an interview Yigal Allon told the author that, together with Yadin and Dayan, he had exerted an indirect influence on the strategy of the 1967 war as an adviser to Prime Minister Eshkol. See also Allon to Liddell Hart 11 October 1967.

35. Interview with Avraham Yoffe. *The Israeli Army* pp. 233, 248, 253.

36. Interview with Ariel Sharon. In insisting that the front commander must interfere in a tactical battle whenever he considered it necessary, he seemed indifferent to the suggestion that this might cause difficulties for the Chief of Staff or the Government. See also *The Israeli Army* pp. 246–247.

37. Ibid. pp. 250–259. *Encounter* op. cit.

38. Dr J. L. Wallach to Liddell Hart 27 June 1967. See also Wallach's

article 'The Israeli Armoured Corps in the Six Day War' in *Armor* LXXVII May–June 1968 pp. 34–43.

39. Laskov to Liddell Hart 22 August 1967. Liddell Hart to Laskov 12 and 25 September 1967.

40. Yadin to Liddell Hart 29 August 1967. Yadin remarked to the author that his reference to Rabin having learnt from Liddell Hart's books should be interpreted in an indirect sense.

41. Although there can be no dispute that the Israelis were taken by surprise, Luttwak and Horowitz (*The Israeli Army* p. 342) argue persuasively that political considerations debarred them from striking first.

42. Professor J. L. Wallach, in an interview on 17 September 1975, remarked that Liddell Hart had been proved correct in warning the Israelis against becoming over-confident.

43. *The Israeli Army* pp. 381–388. Here, in my opinion, the authors are over-critical of Sharon: his decision to cross the canal and keep advancing may have been unorthodox and risky but it was surely paying dividends.

44. 13/80 Liddell Hart's interview with David Kessler published in the *Jewish Chronicle* 24 November 1967; and *Encounter* op. cit. A prominent member of the Israeli Government (1975) expressed wholehearted support for the creation of a Druse state in a letter to Liddell Hart in October 1967.

Conclusion

Like his friend 'Boney' Fuller, Liddell Hart devoted the last years of his life primarily to military history, publishing his *Memoirs* in two volumes in 1965 and completing his monumental *History of the Second World War* just before his death in January 1970. In these years his achievements received belated recognition, and honours and tributes were showered upon him. Senior soldiers, including Field Marshal Montgomery, described him as Britain's outstanding military thinker; German and Israeli generals acknowledged him as their teacher; and for a host of junior officers and scholars he was consulted and admired as a sage.

In 1963 Liddell Hart and Fuller together received Chesney Memorial Medals, awarded for distinguished contributions to military history, at the Royal United Service Institution – an honour which both had earned thirty years earlier. In making the presentation, General Sir John Hackett justly observed that although their 'names are names to conjure with in other countries, [they] have only with the years come to receive true recognition in their own'.

By British social custom all is forgiven to mavericks, rebels, outsiders and critics of the Establishment when they reach the age of seventy. To mark Liddell Hart's seventieth birthday in 1965, Michael Howard edited a *festschrift* entitled *The Theory and Practice of War,* whose distinguished list of contributors included Generals Beaufre and Pile, Brigadier Yigal Allon, Henry Kissinger and Alastair Buchan. That same year Liddell Hart was appointed Distinguished Visiting Professor at the University of California, Davis; and in the New Year's Honours List of 1966 he was knighted. Academic recognition was also forthcoming in the award of an Hon.D.Litt, at Oxford, and an honour which pleased him most of all—particularly as he had not even taken his bachelor's degree—he was made an Honorary Fellow of his old Cambridge College, Corpus Christi.

The complexity of Liddell Hart's endeavours, even within the military field, becomes obvious when one tries to find a neat descriptive label: journalist, commentator, critic, adviser on defence, army reformer, tactician, strategic theorist and—in the popular, non-technical sense—philosopher; he played all these roles at various times and many of them simultaneously. It would be unjust to describe him as the proverbial Jack-of-all-trades because his multifarious pursuits nourished each other, but possibly he would have done even better at some of them, as a historian especially, had he not spread his interests so widely.

Clearly, however, it was not in Liddell Hart's nature to confine himself to specialization in a limited field. He was rather, in Michael Howard's apt description, one of the last of a distinguished line of Sages, whose archetype was the iconoclastic Voltaire and whose English precursors included the Webbs, Wells, Shaw and Russell. Liddell Hart's uniqueness lay in the fact that his chief preoccupation was with warfare, but the description 'military thinker' is patently inadequate:

> The Sage is a monarch, not a member of a republic. Above all the Sage, however deeply his roots may be sunk in the expertise of a single subject, billows uncontrollably outside it. Liddell Hart was no more simply a military thinker than Shaw was simply a playwright or Russell simply a philosopher. Like them, the passionate concern he devoted to his own subject gave him a passionate concern about everything else as well.[1]

Liddell Hart also subscribed to the ideal of the philosopher-king who, having conquered the urge of personal ambition and made himself independent of all political, religious and sectional affiliations, is suited to proffer wise advice to the harassed statesman or general. Despite some obvious imperfections he succeeded in playing this role to a remarkable extent. Apart from his brief association with Hore-Belisha, he held aloof from party politics, and after leaving *The Times* in 1939 he never again had a regular employer. To maintain such a precarious existence was an astonishing feat in terms of industry, endurance and moral integrity considering that he was essentially dependent on his pen to earn his living. In other words, in his egocentric and highly individual way, Liddell Hart was an idealist. This was patently clear to anyone who knew him at all well and comes out strongly in his Papers, particularly in the

1930s. The precise nature of his idealism was unfortunately never fully developed and is largely implicit in his books. He paid a high price in terms of career prospects, wealth and honours in order to maintain his freedom to pursue and publish the truth as he perceived it, but the endeavour brought its own satisfactions. Also, however imperfect its realization, this underlying humanitarian idealism is one of the qualities which should ensure him a special place among those who have devoted their lives to the dispassionate study of war.

What are likely to be the most important and enduring facets of Liddell Hart's work? It seems to me that he was soundest and most directly useful to the military profession as a tactical theorist where he was closest to his own practical experience. At the other end of the spectrum, his philosophy of life, embracing the precepts of willingness to compromise, to limit effort in proportion to the goal, and to practise restraint, moderation and tolerance, provided an admirably humane *leitmotif* to the main corpus of his work. This is not to imply that he did not have much of value to contribute on the middle ground of strategy and policy, but his achievement here is more controversial.

But Liddell Hart's significance in the realm of military thought surely transcends any of the particular strategic theories or doctrines discussed in this book. Two related reasons may be suggested for this claim: the powerful example of a career dedicated to furthering constructive thinking about the nature of war; and the incalculable indirect influence exercised through his help and inspiration of others. In sum, his outstanding achievement, effected by the depth and intensity of a lifetime's effort, may well have been to transform the nature of military thought itself.

Note

1. Michael Howard in *Encounter* April 1970.

Appendix A

Liddell Hart's Principal Publications

New Methods in Infantry Training C.U.P. Cambridge 1918.

The Framework of a Science of Infantry Tactics Hugh Rees, London 1921; revised eds: *A Science of Infantry Tactics* Clowes, London 1923, 1926.

Paris, or the Future of War Kegan Paul, London: Dutton, N.Y. 1925.

A Greater than Napoleon; Scipio Africanus Blackwood, Edinburgh/London 1926: Little, Brown, Boston 1927.

The Remaking of Modern Armies Murray, London: Little, Brown, Boston 1927.

Great Captains Unveiled Blackwood, Edinburgh/London 1927: Little, Brown, Boston 1928.

Reputations: Ten Years After Murray, London: Little, Brown, Boston 1928.

Sherman: Soldier, Realist, American Dodd, Mead, N.Y. 1929; *Sherman: the Genius of the Civil War* Benn, London 1930; Eyre & Spottiswoode, London 1933.

The Decisive Wars of History Bell, London: Little, Brown, Boston 1929; latest ed., with additions: *Strategy: the Indirect Approach* Faber, London: Praeger, N.Y. 1967.

The Real War Faber, London: Little, Brown, Boston 1930; enlarged ed.: *A History of the World War, 1914–1918* Faber 1934: Little, Brown 1935; reissued as *History of the First World War* Cassell, London 1970: Putnam, N.Y. 1971.

Foch: the Man of Orleans Eyre & Spottiswoode, London 1931: Little, Brown, Boston 1932; Penguin, Harmondsworth 1937.

The British Way in Warfare Faber, London 1932: Macmillan, N.Y. 1933; enlarged ed.: *When Britain Goes to War* Faber 1935; with additional chapters and original title: Penguin, Harmondsworth 1942.

The Future of Infantry Faber, London 1933: Military Service Publishing Co., Harrisburg, Pa. 1936.

The Ghost of Napoleon Faber, London 1933: Yale U.P., New Haven 1934

'T. E. Lawrence' in Arabia and After Cape, London 1934: *Colonel Lawrence: the Man Behind the Legend* Dodd, Mead, N.Y. 1934.

The War in Outline, 1914–1918 Faber, London: Random House, N.Y. 1936.

Europe in Arms Faber, London: Random House, N.Y. 1937.

Through the Fog of War Faber, London: Random House, N.Y. 1938.

The Defence of Britain Faber, London: Random House, N.Y. 1939.

Dynamic Defence Faber, London 1940.

The Current of War Hutchinson, London 1941.

This Expanding War Faber, London 1942.

Thoughts on War Faber, London 1944.

Why Don't We Learn from History? Allen & Unwin, London 1944; Hawthorn Books, N.Y. 1972.

The Revolution in Warfare Faber, London 1946: Yale U.P., New Haven 1947.

The Other Side of the Hill Cassell, London 1948: *The German Generals Talk* Morrow, N.Y. 1948; enlarged ed. Cassell 1951; Panther, London 1956.

Defence of the West Cassell, London: Morrow, N.Y. 1950.

The Tanks 2 vols, Cassell, London: Praeger, N.Y. 1959.

Deterrent or Defence Stevens, London 1960: *Deterrent or Defense* Praeger, N.Y. 1960.

Memoirs 2 vols, Cassell, London 1965: Putnam, N.Y. 1965–66.

History of the Second World War Cassell, London: Putnam, N.Y. 1970.

EDITED BY LIDDELL HART

T. E. Lawrence to His Biographer Liddell Hart Doubleday, N.Y. 1938: Faber, London 1939; 2nd ed. (with Robert Graves) *T. E. Lawrence to His Biographers Robert Graves and Liddell Hart* Doubleday: Cassell, London 1963.

The Letters of Private Wheeler, 1809–1828 Michael Joseph, London: Houghton Mifflin, Boston 1951.

The Rommel Papers Collins, London: Harcourt Brace, N.Y. 1953.

The Soviet Army Weidenfeld & Nicolson, London 1956: *The Red Army* Harcourt Brace, N.Y. 1956.

Appendix B

Writings on Liddell Hart: an introductory guide

The only published study fully documented from the Liddell Hart archives is the excellent chapter in Jay Luvaas *The Education of an Army* (U. of Chicago 1964: Cassell 1965), which was emended in detail by its subject. Irving M. Gibson's essay 'Maginot and Liddell Hart' in E. M. Earle (ed.) *Makers of Modern Strategy: Military Thought from Machiavelli to Hitler* (Princeton U.P. 1941: O.U.P. 1944) must be used with caution: as Luvaas comments, it 'reveals as much about the mood and misconception at the time as it does about Liddell Hart, who is viewed here largely through the eyes of his critics'.

The *festschrift, The Theory and Practice of War*, edited by Michael Howard and presented to Liddell Hart on his 70th birthday in 1965 (Cassell 1965: Praeger 1966), contains essays which touch on various aspects of his career, including his influence on the pre-1939 German Army. Robin Higham devotes a good deal of space to Liddell Hart's ideas in *The Military Intellectuals in Britain, 1918–1939* (Rutgers U.P., N.J. 1966). R. H. S. Crossman reprints a perceptive book review entitled 'The Strange Case of Liddell Hart' in *The Charm of Politics* (Hamish Hamilton: Harper 1958). For a critical view of Liddell Hart's association with *The Times* see Donald McLachlan's biography of Barrington-Ward *In the Chair* (Weidenfeld 1971) Chapter 15.

Of the many reappraisals which were occasioned by Liddell Hart's death in January 1970, by far the most interesting was Michael Howard's brilliant essay in *Encounter* (April 1970). His critical review of the Liddell Hart *Memoirs* in *R.U.S.I. Journal* (February 1966) should also be noted; and that of Barry Powers in *Journal of Modern History* (December 1968). Two other obituary articles may be strongly recommended: Ronald Lewin's study in *International Affairs* (January 1971); and Robert O'Neill's in the Australian *Army Journal* (April 1970). Ronald Lewin has also written the entry on Liddell Hart for the forthcoming *Dictionary of National Biography 1961–1970* (O.U.P.). Brian Bond's essay 'Some Nuclear-Age Theories of Sir Basil Liddell Hart' in *Military Review* (August 1970) showed that its subject made important contributions to military thought after 1945 as well as to the *blitzkrieg* era. The only thorough attempt to present a balanced view of Liddell Hart's career and achievements by bringing together the recorded comments of a wide variety of friends, protégés and critics was Robert Pocock's radio programme *Liddell Hart: the Captain who taught Generals*, which was published in condensed form in *The Listener* (28 December 1972).

Stephen Brooks, who organized and catalogued the archives after Liddell Hart's death, has contributed a perceptive essay on 'Liddell Hart and his Papers' in Brian Bond and Ian Roy (eds.) *War and Society Yearbook* II (Croom Helm: Holmes & Meier 1977) pp. 129–40. Typical examples of the rather shallow analyses of Liddell Hart's influence in Israel which the present study has tried to improve upon are: Jac Weller 'Sir Basil Liddell Hart's Disciples in Israel' in *Military Review* (January 1974); and Gary S. Itzkowitz 'Israel's Military Doctrine' in *Times of Israel* (April 1974). Four further foreign assessments to be noted are: Colonel Robert J. Icks 'Liddell Hart: One View' in *Armor* (November–December 1952); Jeffrey A. Gunsberg 'Liddell Hart and Blitzkrieg' ibid. (March–April 1974); Général L. M. Chassin 'Un grand penseur militaire britannique: B. H. Liddell Hart' in *Revue de Défense nationale* (October 1950); and Colonel Ernest Léderrey 'Le capitaine B. H. Liddell Hart' in *Revue Militaire Suisse* (May 1956). Additional short articles, graduate seminar papers and miscellaneous published comments on various aspects of Liddell Hart's career may be found in the Liddell Hart Papers 13/1–5.

Index